9th International Conference on Material Science 2020

IOP Conference Series: Materials Science and Engineering
Volume 1000

Ulaanbaatar, Mongolia
10 - 13 September 2020

ISBN: 978-1-7138-3183-9
ISSN: 1757-8981

Printed from e-media with permission by:

Curran Associates, Inc.
57 Morehouse Lane
Red Hook, NY 12571

CURRAN ASSOCIATES INC. proceedings.com

Some format issues inherent in the e-media version may also appear in this print version.

Institute of Physics
Dirac House, Temple Back
Bristol BS1 6BE UK

Phone: 44 1 17 929 7481
Fax: 44 1 17 920 0979

techtracking@iop.org

Additional copies of this publication are available from:

Curran Associates, Inc.
57 Morehouse Lane
Red Hook, NY 12571 USA
Phone: 845-758-0400
Fax: 845-758-2633
Email: curran@proceedings.com
Web: www.proceedings.com

TABLE OF CONTENTS

IOP Conf. Series: Materials Science and Engineering 1000 (2020) 011001 doi:10.1088/1757-899X/1000/1/011001

Preface

The 9th International Conference on Materials Science (ICMS2020) was held at the National University of Mongolia from September 10 to 13, 2020 in Ulaanbaatar. Unfortunately, the global pandemic had a bad effect on the event. Therefore, some of the participants could not attend the conference in person. A significant part of the participants (more than 50 people) were accommodated in the hall and more than 25 people participated in absentia. Since at this time it is difficult to predict the end of the pandemic, and the termination of the exchange and discussion of scientific results may negatively affect the research of many scientific groups, it was decided to hold this traditional conference using video presentations. Participants from Russia presented their reports and presentations in electronic form (video reports and posters) and were demonstrated at section meetings. In total - about 80 participants.

Organisers were some known scientific organizations:

National University of Mongolia, Ulaanbaatar, Mongolia, National University of Mongolia, Mongolia, Inner Mongolian Normal University, China, Institute of Physical Materials Science, Siberian Branch of the Russian Academy of Sciences, Russia, Mongolian Academy of Sciences, Mongolia, German-Mongolian Institute for Resources and Technology and Erdenet Mining Corporation SOE, Mongolia.

Scientific committee was consisted of famous scientists:

Prof. J.Davaasambuu (National University of Mongolia)

Prof. U.Pietsch (University of Siegen, Germany)

Prof. G.Eckold (University of Goettingen, Germany)

Prof. O.Tegus (Inner Mongolian Normal University, China)

Dr. A.Nomoev (IPMS SB RUS, Russia)

Prof. S.Gorfman (Tel Aviv University, Israel)

Prof. S.Bardakhanov (Buryat State University, Russia)

Prof. P.Daniel (Le-Mans University, France)

Prof. B.Battsengel (GMIT, Mongolia)

Dr. B.B.Damdinov (SFU, Krasnoyarsk, Russia)

Acad. O.Penyazkov (A.V.Lykov Institute, Belarus)

Acad. D.Sangaa (Mongolian Academy of Sciences)

Acad. J.Temuujin (Mongolian Academy of Sciences)

Dr. V.Koledov (IRE, RAS, Russia)

Dr. P.Altantsog (Mongolian Academy of Sciences

IOP Conf. Series: Materials Science and Engineering 1000 (2020) 011001 doi:10.1088/1757-899X/1000/1/011001

All materials were reviewed by the editors: Dr. N. Tuvjargal and Prof. J.Davaasambuu from Department of Physics, Faculty of Arts and Sciences, National University of Mongolia.

The conference consisted of plenary presentations for 30 minutes and session reports for 15 minutes. There were 35 oral reports and 29 posters in total. The organizers have provided 5 minutes for questions that can be asked in person or via the Internet.

The objectives of the conference were to bring together scientists working in the field of materials science from around the world, to exchange ideas, recent research results in this area and to offer a platform for the initiation of scientific cooperation between Mongolian and foreign scientists. The conference was focused on the structures, properties and applications of new materials as well as their characterization techniques.

Topics were: Materials science, Condensed matter physics, Computational methods, Experimental methods and instrumentation, Laser spectroscopy.

At the conference, nanotechnology-related studies related to quantum effects, such as carrier tunneling, controlled changes in the structure of energy bands, quantization of energy spectra of charge carriers and spin phenomena, and the study of new physical features of semiconductor nanostructures, were discussed. The topic were a description of the transport properties of various ion-exchange membranes, crystallization of supercooled liquids and amorphous solids, increase the corrosion resistance of aluminum alloys, obtaining nanoscale metal and alloy powders. Numerical simulation of phonon crystals with time-dependent properties of a periodic structure, computer simulation by the molecular dynamics method, the processes of structure formation of nanoclusters of an alloy of silver and copper were presented.

The organizers and participants of ICMS 2020 noted the high scientific level of the presented reports.

Peer review declaration ICMS2020

All papers published in this volume of *IOP Conference Series: Materials Science and Engineering* have been peer reviewed through processes administered by the Editors. Reviews were conducted by expert referees to the professional and scientific standards expected of a proceedings journal published by IOP Publishing.

- **Type of peer review: Single-blind**

Single-anonymous: reviewers' identities are hidden from authors

- **Describe criteria used by Reviewers when accepting/declining papers. Was there the opportunity to resubmit articles after revisions?**

Criteria used: novelty, motivation, research quality, presentation quality, discussion of the context of the work,

Authors had the opportunity to resubmit articles after revision.

- **Conference submission management system: Direct mailing to the conference secretary**
- **Number of submissions received: 82**
- **Number of submissions sent for review: 11**
- **Number of submissions accepted: 11**
- **Acceptance Rate (Number of Submissions Accepted / Number of Submissions Received X 100): 100**
- **Average number of reviews per paper: 1**
- **Total number of reviewers involved: 3**
- **Any additional info on review process (ie plagiarism check system): plagiarism check system was used.**
- **Contact person for queries: A.V. Nomoev**

IOP Conf. Series: Materials Science and Engineering 1000 (2020) 012001 doi:10.1088/1757-899X/1000/1/012001

Structural, surface and optical properties of nanoalumina produced by various ways

V V Syzrantsev[1(a)], T V Larina[2], Yu A Abzaev[3], E A Paukstis[2], A I Kostyukov[4]

[1] Institute of Materials Science of the Siberian Branch of the Russian Academy of Sciences, Ulan-Ude, Sakhyanovoy str. 6, 670070, Russia;
[2] Institute of Catalysis of the Siberian Branch of the Russian Academy of Sciences, Lavrentieva av. 5, Novosibirsk, 630090, Russia;
[3]Tomsk State University of Architecture and Building, Tomsk, Solyanaya sq. 2, 63400, Russia 3;
[4] Novosibirsk State University, Pirogova Str. 2, 630090, Novosibirsk, Russia.

[(a)]vvveliga@mail.ru

Abstract. Start The compared study of the properties of liquid-phase and electric explosion alumina nanoparticles was made. The X-ray diffraction analysis demonstrated that these nanoparticles remain in amorphous and semi-amorphous states. The complete structural information of Al_2O_3 nanoparticles was obtained as a result of the full-profile refinement of their model phase parameters. The parameters of the unit cells, spatial distribution of atoms, and occupancy of nodes were also determined.The data of IR spectra of surface OH groups and adsorbed CO revealed that the surface imperfection of an electric explosive sample is lower than that obtained using the liquid-phase method. Using photoluminescent spectroscopy, impurities of Cr^{3+}, Mn^{4+} and Ti^{3+} in the octahedral oxygen coordination and Fe^{3+} in the tetrahedral oxygen coordination were detected. It was observed that the concentration of these impurities was significantly higher in the electric explosion sample. Using UV-Vis DR spectroscopy, it was found that the liquid-phase sample was amorphous, and the electric explosive sample was well crystallized and most probably consisted of 2D and 3D nanostructures.

1. Introduction

The synthesis modification of nanoparticle properties is an interesting and practically relevant task. Nanoparticles often do not have a crystalline structure (are X-ray amorphous) or the content of the crystalline phase is very low and can not be identified using conventional X-ray phase analysis methods. The properties of nanoparticle surfaces are the most important factors for determining the efficiency of their application. If the surface groups coincide, their heterogeneity becomes very important. For example, the ratio of the bridging to terminal –OH groups [1] and the number of Lewis acid sites [2] can vary. In addition, the synthesis method can introduce various impurities that also have a significant effect on the properties of nanoparticles. This leads to a variation of the catalytic activity, the intensity of the interaction of nanoparticles with the environment, particularly in ceramics [3], polymers [4, 5] etc. A similar effect was observed for alumina nanoparticles in [6], in which a similar comparative study of the structural state of nanoparticles obtained by electron beam evaporation and high-temperature hydrolysis was performed.

This work aims to identify and provide a model description of the structural state of Al_2O_3 nanoparticles synthesised by electric explosion and liquid-phase methods; determination of the quantitative phase composition, structure parameters. Additionally, comparison of the surface characteristics of the samples and their relationship with the structural properties of the selected nanoparticles was performed.

2. Materials and methods

We used alumina nanoparticles synthesized by the liquid-phase method (LP) (China; mean size 25 nm) and the electric explosion method (mean size 55 nm) (EE) [7] in our study. The size of the initial nanoparticles was measured using TEM (JEOL-2010, Japan).

X-ray diffraction studies of Al_2O_3 were carried out on DRON4-07. The surveys were performed on copper radiation (Kα), according to the Bragg-Bretano scheme, with a step of 0.020. The exposure time at a point was 1 s in the angular range of 18–86^0. The voltage on the X-ray tube was 30 kV and the beam current was 25 mA. A quantitative phase analysis (QPA) of Al_2O_3 was carried out based on the Rietveld method [8]. As standards, we used lattices from the crystallographic bases COD and AMCSD [9, 10], (No. 96-120-0006, Corundum_0012922). The standards are symbolically called T and S, respectively. The X-ray diffraction analysis revealed that, as in [6, 8], alumina samples substantially had the same phases. Interestingly, the real states of the alumina samples were not strictly in the α phase, but all of them were in the X-ray amorphous state. Details of the X-ray studies of the corresponding "semi-amorphous" states can be found in[8].

The IR spectroscopy samples were pressed into tablets with a thickness (ratio of tablet weight to its geometric area) of 15 to 49 mg / cm^2. The tablets were placed in a quartz IR cuvette and evaluated for 1 hour (vacuum of 10^{-5} Pa). After this, the samples, without coming in contact with the atmosphere, were transferred to a measuring chamber which allowed the recording of IR spectra at temperatures up to 77 K. Subsequently, we took the IR spectra of the samples before adsorption of CO at room temperature and then at 77–110 K. The IR spectra were taken from the Shimadzu IRTraser-100 spectrometer in the region of 600–6000 cm^{-1} with a resolution of 4 cm^{-1}. The IR spectra were accumulated through 100 scans which ensured a signal-to-noise ratio of at least 75000. The adsorption of CO was carried out at 77 K with doses that ensured pressure values in the cuvette as 0.1, 0.4, 0.9, 1.4 and 10 Torr.

The optical properties of the samples were studied using Ultraviolet-Visible Diffuse Reflection Spectroscopy (UV-Vis DRS) using a Shimadzu UV-2501 PC spectrophotometer with an ISR-240A diffuse reflection attachment. The samples in the form of a powder were placed in a quartz cuvette with an optical path length of 2 mm. The spectra were recorded relative to the reflection standard - $BaSO_4$ in the range 190–900 nm (11000–54000 cm $^{-1}$) with a step of 2 nm. Reflection coefficients were converted to absorption coefficients using the Kubelka-Munk function $F(R)=(1-R)^2/2R$, in which R is the diffuse reflection coefficient depending on the wave number expressed in cm^{-1}. The obtained UV-Vis DRS data are presented in coordinates: the Kubelka–Munk function F(R) vs the wave number.

The photoluminescence (PL) spectra were measured using a Cary Eclipse (Varian) spectrofluorimeter in the wavelength range of 550-900 nm, with selective excitation of the wavelength region which is significant for the centre under study. Samples were placed in a quartz cuvette with an optical path length of 2 mm, and irradiation and collection of luminescence were carried out in the "reflection" mode. The measurements were performed under comparable conditions to compare the parameters of the observed PL. The PL excitation (PLE) spectra were recorded in the region of 200–600 nm. Processing of the PL spectra was performed in the OriginPro 15 program.

3. Results and discussion

3.1. X-ray modeling

Figure 1 shows the spatial distribution of atoms and the results of a full-profile refinement by the Rietveld method. Figure 2 shows the experimental diffraction patterns (a), b) curve 1), integrated calculated intensities (curve 2), their difference (curve 3), and the calculated intensities of the individual phases in LP synthesis. For electric explosion synthesis, the refinement was carried out by the lattice T, whose contribution to the intensity turned out to be dominant (Fig. 1a).

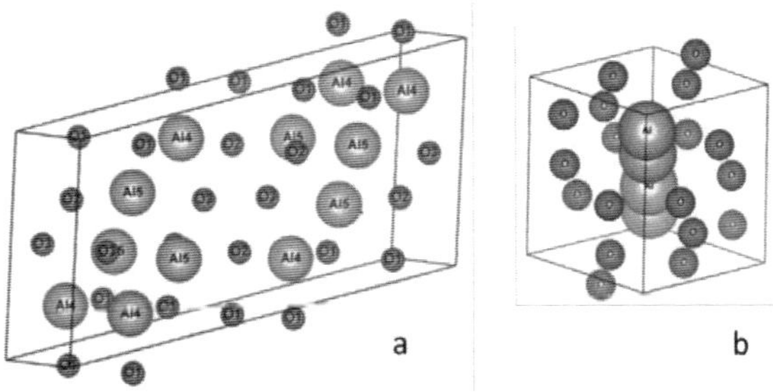

Figure 1: Spatial distribution of atoms: lattice T (a); lattice S (b).

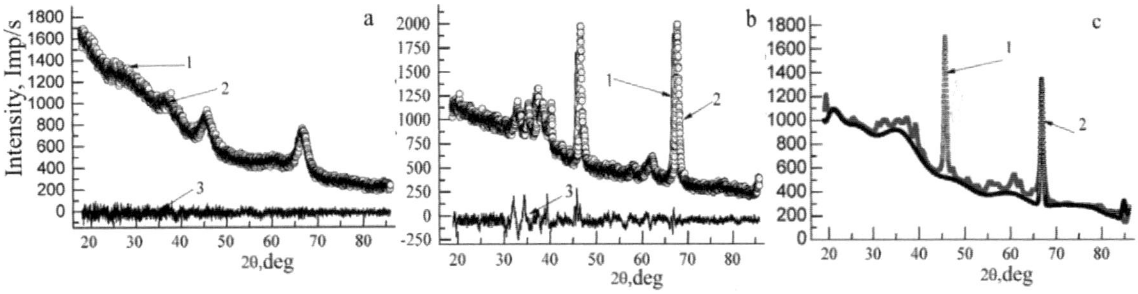

Figure 2: Diffraction patterns of the alumina nanoparticles. Synthesis: a) LP method; b) EE method. 1 - experimental, 2 - theoretical diffractograms, 3 - intensity difference. c) The calculated intensities of the phases: 1 - T; 2 - S.

Tables 1–4 show the structural parameters of lattices T and S in the initial state as well as after full-profile refinement. The tables also show the agreement criteria (R_{wp}) and fraction of the "explained" intensity of the experimental diffractograms. According to the results obtained (R_{wp} criterion, intensity difference, Fig. 2), the theoretical intensities and quantitative phase contents are in accordance with the experimental ones. The acceptance criterion was found to be equal: R_{wp} = 3.96, 7.37%, respectively. The structural parameters of the lattices of these phases are given in Tables 1. Tables 2, 3, and 4 show the relative coordinates of the atoms of the standards (initial) and the atoms after a full-profile refinement.

Table 1. Alumina structure parameters.

Structure		a [nm]	b [nm]	c [nm]	α [deg.]	β [deg.]	γ [deg.]	Singony
T (LP)	Initial	1.1795	0.2910	0.5621	90.00	103.79	90.00	C2/m, monoclin.
	Refinement	1.20475	0.27032	0.6107	90.00	104.71	90.00	
T (EE)	Initial	1.17950	0.29100	0.5621	90.00	103.79	90.00	C2/m, monocl.
	Refinement	1.19473	0.28528	0.5539	90.00	105.48	90.00	
S (EE)	Initial	0.5120	0.5120	0.5120	55.28	55.28	55.28	R-3C, trigon
	Refinement	0.51236	0.51236	0.51236	55.03	55.03	55.03	

Table 2. The relative coordinates of atoms in the lattice T (LP).

Atoms		x	y	z	Equval.	Occup.	R_{wp}	Part, [%]
O	Initial	0.16619	0	0.09877	0.01267	1.00	3.964	97.152
	Refinement	0.163	0	0.123	0.01267	0.01267		
O	Initial	0.56966	0	0.31235	0.01267	1.00		
	Refinement	0.489	0	0.261	0.01267	0.01267		
O	Initial	0.91617	0	0.46845	0.01267	1.00		
	Refinement	0.83	0	0.439	0.01267	0.01267		
Al	Initial	0.15117	0	0.83127	0.01267	1.00		
	Refinement	0.101	0	0.794	0.01267	0.01267		
Al	Initial	0.27202	0	0.64758	0.01267	1.00		
	Refinement	0.352	0	0.687	0.01267	0.01267		

Table 3. The relative coordinates of atoms in the lattice T (EE).

Atoms		x	y	z	Equval.	Occup.	Part, [%]
O	Initial	0.163	0	0.123	0.01267	1	79.134
	Refinement	0.05206	0	0.00074	0.01267	1	
O	Initial	0.489	0	0.261	0.01267	1	
	Refinement	0.56517	0	0.27674	0.01267	1	
O	Initial	0.83	0	0.439	0.01267	1	
	Refinement	0.94956	0	0.59524	0.01267	1	
Al	Initial	0.101	0	0.794	0.01267	1	
	Refinement	0.14416	0	0.81285	0.01267	1	
Al	Initial	0.352	0	0.687	0.01267	1	
	Refinement	0.26246	0	0.63903	0.01267	1	

Table 4. The relative coordinates of atoms in the lattice S(EE).

Atoms	x	y	z	Equval.	Occup.	Rwp	Part, [%]
O	0.553	-0.053	0.25	0	1	7.369	17.028
Al	0.355	0.355	0.355	0	1		

It can be assumed that

1) the main fraction of the LP particles was the amorphous phase (~ 0.97 T, Table 2),

2) the main fraction of the EE particles was the superposition of the phases T and S (Table 3, 4), the proportion of which was ~ 0.79 and 0.17, respectively.

Figure 2 shows that the diffraction pattern of the alumina synthesised by the LP method corresponds to an amorphous state. In fact, the role of the main reflexes in the integrated intensity is insignificant (Fig. 2a. curve 1). The X-ray amorphous state is formed by 12 polyhedra with a central Al atom. Polyhedra can be of two types: angular and with edges [11]. The semi-amorphous state of electric explosion nanoparticles is characterised by six polyhedra (with edges). We estimated the mixing energy to analyse the stability of lattices with refined structural parameters. Details of the calculations are given in [12]. Modeling was carried out from the first principles by the formula $E_c = E_t - [xE_{Al} + yE_O]$, where E_t, E_{Al}, E_O are the energy of the Al_2O_3 lattice and of pure Al and O elements, respectively. The x, y values are the relative number of O and Al elements in Al_2O_3, which are equal for T and S lattices: $x = (\alpha / 4)$, $y = (\beta / 4)$. The numbers (α, β) are (8, 12) and (4, 6) for the lattices T and S, respectively.

The binding energy of atoms in the lattices is $|E_c|$. The results of the mixing energy calculations are given in Table 5, which also shows the energies of the lattices of pure elements. As follows from the above results, the alumina mixing energies in different states turn out to be substantially negative, i.e. model lattices are in highly stable atomic configurations [8].

Table 5. The mixing energy of Al_2O_3

Method	Phase	E_t, [eV]	E_c, [eV]	E_{Al}, [eV]	E_O, [eV]
Liquid phase	T	-5723.625753	-356.89	-228.63489	-3272.9775
Electric explosion	T	-5649.928373	-279.799		
	S	-2863.38232	-184.906		

Additionally, we calculated the elastic moduli of Al_2O_3 using GULP [13] for amorphous and semi-amorphous lattices T and S after preliminary geometric optimisation using the Streitz-Mintmire potential [14]. The calculation results are provided in Table 6. For comparison, Table 6 presents the experimental values of the elastic characteristics of crystalline phases [11]. The analysis showed that the elastic moduli of amorphous and semi-amorphous phases are lower than the corresponding values for crystalline phases.

Table 6. Elastic characteristics of Al_2O_3

Method	Phase	B [GPa]	G [GPa]	E [GPa]
Liquid phase	T	192.384	102.335	341.865
Electric explosion	T	192.384	102.335	340.952
	S	241.174	137.823	358.564
Solid	γ	253	164	404.6
Solid	α	192	102	342.0

3.2. IR spectroscopy

Fig. 3a presents the IR spectra for a sample obtained using LP synthesis. It consists of five bands of OH groups: 3795, 3780, 3732, 3695, and 3670 cm^{-1}. The first two bands are related to terminal OH groups associated with three- and five- coordinated (defective) aluminum ions. The remaining bands belong to OH bridging groups connected simultaneously with two aluminum ions in different coordination. Due to the broadening of the bands, it was very difficult to make a detailed interpretation of the spectra. According to the strength of the acid centres, the hydroxyl groups of 3795 cm^{-1} were very weak, because their shift during CO adsorption was 30 cm^{-1}.

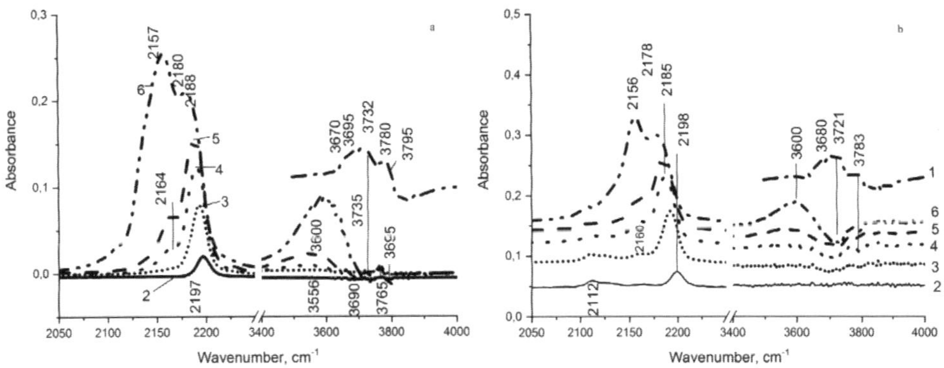

Figure 3: IR spectra of CO adsorbed on LP alumina (a) and EE alumina (b). 1 - IR spectrum of OH groups of the ignite sample (thickness 35 mg / cm^2), 2-6 IR spectra recorded after CO adsorption at pressures of 0.1; 0.4; 0.9; 1.4 and 10 Torr.

The bands related to bridging OH groups upon CO adsorption show the shift of 135–140 cm^{-1}, which implies that these are weakly acidic OH groups. In the region of CO valence vibrations (2050-2250 cm^{-1}), bands 2197–2180 cm^{-1} are observed, which relate to complexes of CO with Lewis acid centres of medium strength. The 2164 cm^{-1} band belong to CO complexes with OH groups of medium acidity, but the corresponding OH groups, against the background of weaker complexes (band 2157 cm^{-1}) with bridging OH groups, cannot be detected. Probably the band 3690 cm^{-1} can be attributed to these OH groups.

The EE sample IR spectra of the OH groups (Fig. 3b) shows two bands 3783, 3721, and a shoulder of approximately 3680 cm^{-1}. The first band refers to terminal OH groups associated with 3 and 5 coordinated (defective) aluminum ions; probably due to the broadening of the bands, 3780 cm^{-1} and 3795 cm^{-1} merge into one range. The remaining bands belong to OH bridging groups connected simultaneously with two aluminum ions in different coordination. The terminal OH groups, under CO adsorption (Fig. 3b), showed more than 30 cm^{-1} shift, which implies that these OH groups have stronger acidity. The bridging OH groups were weaker than for the LP sample, as the shift during CO adsorption was about 120 cm^{-1}, instead of 135 cm^{-1}. The strongest Lewis centres are characterized by a band of 2197 cm^{-1}.

In the region of CO valence vibrations (2050–2250 cm^{-1}), EE sample cleared bands of 2198–2178 cm^{-1}, which were related to complexes of CO with Lewis acid centres of medium strength. The strongest Lewis centres were found in the band 2189 cm^{-1}. Additionally, in the IR spectrum upon adsorption of the first portion of CO, a band of 2112 cm^{-1} was cleared, which can be attributed to the adsorption of CO on Ti^{3+} clusters.

Therefore, it can be concluded that the LP sample was quite close to the well-crystallized α-phase of alumina with a slightly pronounced increased defectiveness (the relative intensity of the band is 3780-3795 cm^{-1} compared to bands below 3735 cm^{-1}) than for well crystallized alumina. This is probably due to a higher ignition temperature. A similar result was obtained from the data on the properties of the Lewis centres. The EE sample probably contained a small amount of titanium oxide (concentrations about of 1-2 mkmol / g (0.006-0.01% wt.)), due to the technological aspects of its synthesis way.

3.3. PL spectroscopy

It is known that, regardless of the synthesis method, defects in the form of anionic oxygen vacancies and 3d impurity elements are always present in Al2O3 [15–18]. In the PL study, the focus was on the luminescence of 3d elements, due to their potential use as structurally sensitive probes. The most commonly analyzed impurities in Al2O3 are Cr3+ ions. The studies demonstrated that upon selective excitation with λ = 532 nm in the region characteristic of the 4A2 → 4T2 electronic vibrational transition of Cr3+ ions, the PL of both samples appears in the region of 650-850 nm with a maximum at 693 nm (Fig. 4). This band results from the inhomogeneously broadened 2E → 4A2 electronic transition in the octahedrally coordinated Cr3+Oh ions. The FWHM value for the LP sample is larger which can be associated with the smaller size of nanoparticles. For both samples, the PL is represented as a complex curve. In addition to the main maximum, the PL spectra show shoulders with maxima at 676, 687 and 710 nm. The EE sample spectra show shoulders with peaks at ~ 750 nm and ~ 800 nm and also present bands with maxima of 705 and 713 nm.

The complex shape of the observed PL is due to the manifestation of various impurities of 3d elements, the luminescence of which can also occur at λ_{ex} = 532 nm. The PL band with λ_{max} = 676 nm is due to the inhomogeneously broadened electronic transition $^2E \rightarrow {}^4A_2$ in octahedrally coordinated $Mn^{4+}{}_{Oh}$ ions. The wider bands with maxima at ~ 750 and ~ 800 nm results from luminescence by tetrahedrally coordinated $Fe^{3+}{}_{Td}$ ions ($^4T_1 \rightarrow {}^6A_1$ transition) and octahedrally coordinated $Ti^{3+}{}_{Oh}$ ions (transition $^2E \rightarrow {}^2T_2$), respectively [15-18].

In the case of Cr^{3+}, its PL depends on the local environment; therefore, the observed bands with λ_{max} = 686 nm, λ_{max} = 693 nm, λ_{max} = 702 nm and λ_{max} = 713 nm can correspond to the emission of

Cr^{3+} in the structure of alumina, whose local structure corresponds to various polymorphic modifications of Al_2O_3. The position of the band with λ_{max} = 693 nm was very close to the known value for Cr^{3+} in high-temperature α-Al_2O_3, with λ_{max} = 687 nm for Cr^{3+} in θ-Al_2O_3 and with λ_{max} = 702 nm for Cr^{3+} in γ-Al_2O_3 and / or δ-Al_2O_3 [15-19]. Moreover, the bands at 676, 705, and 714 nm correspond to vibronic sidebands in Cr^{3+} ions which would then prove the presence of a local α-Al_2O_3 structure in the EE sample (Fig. 4, curve 2).

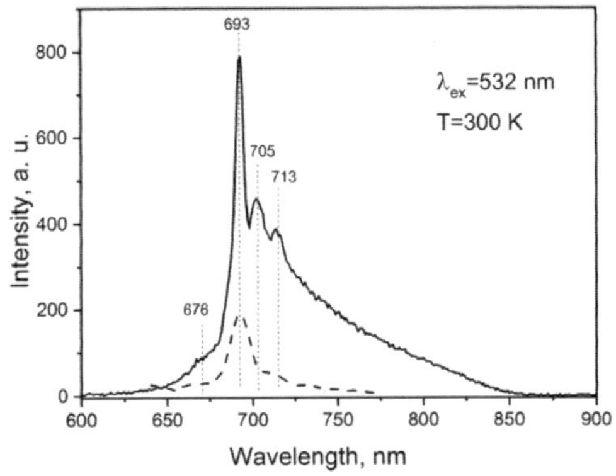

Fig. 4: The PL spectra of samples obtained: 1 - by the LP method (dashed curve, signal intensity increased by 25 times); 2 - by EE method (solid curve), λ_{ex} = 532 nm.

The PL decay kinetics for the band with λ_{max} = 693 nm ($^2E \rightarrow {}^4A_2$ transition in Cr^{3+}) lied in the millisecond interval and is better described by a biexponential dependence with a fast τ_1 = 0.6 ms and a slow component τ_2 = 4.2 ms.

The PLE spectrum of Cr^{3+} ions in Al_2O_3 (λ_{max} = 693 nm, $^2E \rightarrow {}^4A_2$) recorded for the EE sample (Fig. 5) contained intense broad bands in the regions 350 - 475 nm and 500 - 625 nm. These bands with maxima at λ_{max} = 405 nm and λ_{max} = 562 nm correspond to transitions $^4T_1 \rightarrow {}^4A_2$, $^4T_2 \rightarrow {}^4A_2$, respectively.

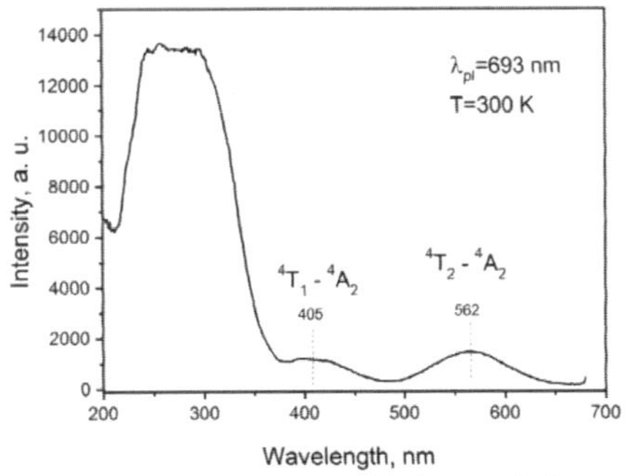

Figure 5: PLE spectrum of the EE sample, λ_{PL} = 693 nm.

The maxima of these bands were very close to the known data for Cr^{3+} in α-Al_2O_3. In the UV region of the spectrum, there was a wide band in the region of 200-370 nm, which does not apply only to Cr^{3+} in α-Al_2O_3. Due to the considerable overlap of wide PL bands from different luminescence centres, the PLE spectra in the region of 200–370 nm recorded for the band λ_{max} = 693 nm, could contain luminescence centres other than Cr^{3+}. Of the most frequently encountered ions in Al_2O_3 that can produce a band in this region, we could distinguish $Fe^{3+}{}_{Td}$, $Ti^{3+}{}_{Oh}$, and $Mn^{4+}{}_{Oh}$ ions. The PLE bands in this region for them, as a rule, were due to the ligand-to-metal charge-transfer bands (LMCT). Excitation of an intense band in the UV spectral region with λ_{ex} = 310 nm made it possible to obtain intense emission bands in the wavelength range 650–875 nm, with a maximum near 750 nm and a shoulder with a maximum at ~ 800 nm (Fig. 6).

The bands with λ_{max} ~ 747 nm and λ_{max} ~ 800 nm were reproduced perfectly when the initial spectrum decomposed into two Gaussian components. Moreover, in the range 650–700 nm for a liquid-phase sample, under such excitation, a less intense band with a maximum at 676 nm appeared in the shoulder. Along with the analysis of the PL bands of these samples at λ_{ex} = 532 nm, the PL bands at ~ 750 nm and ~ 800 nm in Fig. 10 can be unambiguously attributed to the PL of tetrahedrally coordinated $Fe^{3+}{}_{Td}$ ions and octahedrally coordinated $Ti^{3+}{}_{Oh}$ ions and at ~ 676 nm to octahedrally coordinated $Mn^{4+}{}_{Oh}$ ions. As in the case of $Cr^{3+}{}_{Oh}$ ions, the intensity of the PL bands of $Ti^{3+}{}_{Oh}$ and $Fe^{3+}{}_{Td}$ in the electric explosion sample (curve 2) was significantly higher, which indicates a high concentration of these impurities in this sample; this is consistent with the DRS data.

The PL decay kinetics of an electroexplosive sample for a band with λ_{max} = 750 nm ($^4T_1 \rightarrow {}^6A_1$ transition in $Fe^{3+}{}_{Td}$) at λ_{ex} = 532 nm, as in the case of $Cr^{3+}{}_{Oh}$ ions, lies in the millisecond range and is better described by a biexponential dependence with fast τ_1 = 1.0 ms and the slow component τ_2 = 4.7 ms. The PLE spectrum for this band contains only an intense wide band in the UV region of the spectrum with a maximum at ~ 300 n; this results from the O^{2-} - Fe^{3+} charge-transfer band and also through the partial contribution of O^{2-} - Ti^{3+} due to the strong overlap of the PL bands with ~ 750 nm and ~ 800 nm.

3.4. UV-Vis DRS results

UV-Vis DRS spectrum shows that the LP sample (Fig. 7a) clears three absorption regions: up to 20,000 cm^{-1}, 20,000–40,000 cm^{-1} and a region above 40,000 cm^{-1}. The nature of the UV-Vis DRS spectrum of the EE sample (Fig. 7b) was completely different: absorption was observed below 25,000 cm^{-1}, a wide shoulder 30,000–42,000 cm^{-1}, and absorption in the region above 42,000 cm^{-1}.

It is known [20] that in the IR region a weak absorption band results from the d–d transition of Fe^{2+} cations in octahedral oxygen coordination for which LMCT bands appear in the UV spectral region. In oxygen-containing matrices, such $Fe^{2+}{}_{Oh}$ cations are isolated and embedded in the lattice of this matrix and produce $FeAl_2O_4$ type clusters. The reason is that the Al_2O_3 matrix can fix only $Me^{3+}{}_{Oh}$, such as Cr^{3+}, Fe^{3+} and Ti^{3+}cations due to an isomorphic substitution of Al^{3+} cations in the lattice. Furthermore we discovered them in these samples using PL spectroscopy.

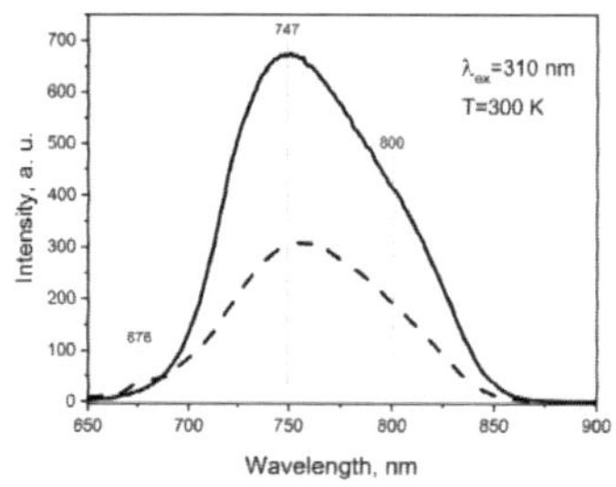

Figure 6: The PL spectra of 1 – the LP sample (dashed curve); 2 - the EE sample (solid curve, the signal intensity was reduced 93 times), λ_{ex} = 310 nm.

Based on the analysis of the UV-Vis DRS spectrum of a LP sample, it was observed that absorption of about 30,000 cm^{-1} has a lower intensity than absorption in the region up to 20,000cm^{-1}. This indicates that this sample contains impurity ions absorbing in the near IR region at a border with the visible region, such as $Ti^{3+}{}_{Oh}$ and $Fe^{2+}{}_{Oh}$, almost twice as much as ions absorbing only in the visible region at the border of the UV region of the absorption spectrum such as $Cr^{3+}{}_{Oh}$ and $Mn^{4+}{}_{Oh}$. In this case, the UV region of the UV-Vis DRS spectrum above 40,000 cm^{-1} shows the LMCT bands of the mentioned d elements, which were earlier detected and uniquely identified by PL spectroscopy. EE sample (Fig. 7) presents a completely different picture of the UV-Vis DRS spectrum. The visible region shows up, as in the case of a LP sample, d-d transitions of impurity d elements, such as $Ti^{3+}{}_{Oh}$ and $Fe^{2+}{}_{Oh}$, most likely contained in a feedstock wire.

Based on the analysis of absorption in the region above 30,000 cm^{-1} in the UV-Vis DRS spectrum, it can be suggested that the EE sample consisted of two parts. The first part is well crystallized (3D structure) which absorbed in the UV region of the UV-Vis DRS spectrum above 42,000 cm^{-1}. The second part is 2D crystalline structures, similar to the alumina obtained by microwave activation of the initial gibbsite $Al(OH)_3$ [21].

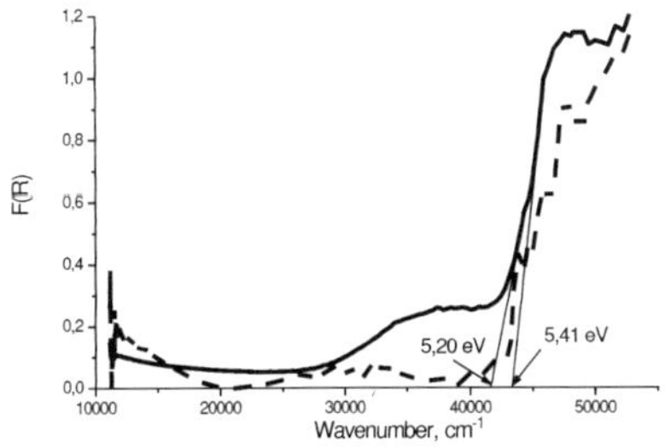

Figure 7: UV-Vis DRS spectra of: a) the LP sample (dush, the intensity is increased 100 times); b) the EE sample (solid).

Moreover, the total absorption in the entire range for the LP sample was almost 100 times less than that for the EE sample. This indicates that the LP sample was amorphous, and the EE sample was well crystallized and probably consisted of 2D and 3D aluminum-containing structures, similar to the alumina samples obtained through microwave irradiation of the initial gibbsite $Al(OH)_3$ [21]. In this case, the LP sample contained the main $Fe^{2+}{}_{Oh}$ cations. The content of $Ti^{3+}{}_{Oh}$ cations and other impurity cations ($Mn^{4+}{}_{Oh}$, $Fe^{3+}{}_{Td}$ and $Cr^{3+}{}_{Oh}$) was low, but was discovered using PL spectroscopy. However, the EE sample contained only $Fe^{2+}{}_{Oh}$ cations.

Thus, it was shown that the surface and optical properties of nanoparticles can unpredictably differ depending on the method of their synthesis, despite the proximity of their structure. A comparison with the results obtained in [6] confirms this conclusion and necessitates a more detailed study of the properties of nanoparticles before their practical use.

4. Conclusion

The structural state of alumina produced by the LP and EE methods was examined using the Rietveld and simulation methods. The Rietveld method demonstrated that an amorphous state was formed in the first case, with T phase being the main part phase. In the second case, alumina nanoparticles consisted of a mixture of phases based on the lattices T and S, with T being the dominant structure. The distribution of atoms in polyhedra and other types as well as a variation in the parameters of model lattices were demonstrated. The distinguishing features of the structural state were the differences in the total and binding energies. Due to the higher binding energy, the alumina synthesised by the LP method was more stable.

Using IR spectroscopy, it was demonstrated that the fraction of terminal OH groups and their strength on the surface of EE sample was lower than that of LP sample.

The PL of impurity 3d elements revealed that the concentration of impurities was significantly higher in the EE sample, which is associated with the features of the synthesis. The analysis of the PL, PLE spectra, and PL decay kinetics allowed the determination of the coordination of the detected impurity ions. The use of $Cr^{3+}{}_{Oh}$ ions as structurally sensitive probes allowed to reveal a local structure close to well crystallized α- phase in both samples.

The UV-Vis DRS cleared that the EE sample contained well-crystallized, probably, 2D and 3D Al_2O_3 nanostructures, and the LP sample was amorphous. Moreover, the LP sample contained significantly less impurity cations and had almost a 100-fold lower absorption compared to the EE sample.

Acknowledgments

The studies of UV-Vis DR, IR and PL spectroscopy of the Al_2O_3 samples are conducted using the equipment of the Center of Collective Use "National Center of Catalyst Research".

This work was supported by Ministry of Science and Higher Education of the Russian Federation (project No. AAAA-A17-117041110045-9 and project No. 0336-2019-0008).

References

[1] Lamberov A A , Romanova R G , Shmelev I G and Sopin V F 2002 *Rus. J. App. Chem.* **75** 396
[2] Morterra C, Bolis V and Magnacca G 1994 *Langmuir* **10** 1812
[3] Guo R –F, Shen P, Zang J, Ma Y-H and Jiang Q-C 2019 *Ceramics Intern.* **45** 24031
[4] Fu S, Sun Z, Huang P, Li Y and Hu N 2019 *NanoMaterial Science* **1** 12
[5] Syzrantsev V V, Zobov K V, Zavjalov A P and Bardakhanov S P 2015 *Doklady Physics* **60** 46
[6] Syzrantsev V V, Nomoev A V, Bardakhanov S P, Paukshtis E A, Larina T V and Chesalov Y A 2018 *J. of Nanomaterials* 2065687
[7] Lerner M I, Svarovskaya N V, Psakhie S G and Bakina O V 2009 *Nanotech. in Russia* **4** 741
[8] Abzaev Yu A, Syzrantsev V V and Bardakhanov S P 2017 *Rus. Phys. J.* **60** 522
[9] http://www.crystallography.net/search.html
[10] http://www.rruff.geo.arizona.edu/AMS/minerals/Corundum
[11] Davis S and Gutierrez G (2011) *J. Phys.: Condens. Matter.* **23** 495401.
[12] Abzaev Yu A, Sarkisov U S , Afanas'ev D A, Klopotov A A, Gorlenko N and Klopotov V D (2014) *Adv. Materials Research* **1013** 102
[13] Gale J D, Rohl A L (2003) *Molec. Simulation* **29** 291
[14] Streitz F H and Mintmire J W 1994 *Phys. Rev. B* **50** 11996
[15] Kostyukov A, Baronskiy M, Rastorguev A, Snytnikov V, Snytnikov Vl, Zhuzhgov A and Ishchenko A 2016 *RSC Adv.* **6** 2072
[16] Rastorguev A, Baronskiy M, Zhuzhgov A, Kostyukov A, Krivoruchko O, Snytnikov V 2015 *RSC Adv.* **5** 5686
[17] Baronskiy M, Rastorguev A, Zhuzhgov A, Kostyukov A, Krivoruchko O and Snytnikov V 2016 *Opt. Mat.* **53** 87
[18] Kostyukov A I, Zhuzhgov A V, Kaichev V V, Rastorguev A A, Snytnikov Vl N. and Snytnikov V N 2018 *Opt. Mat.* **75** 757
[19] Baronskiy M G, Kostyukov A I, Larina TV, Snytnikov V N, Zaitseva N A and Zhuzhgov A V 2019 *Mat. Chem. Phys.* **234** 403
[20] Lever A B .P Inorganic Electronic Spectroscopy (Elsevier: Amsterdam – Oxford – New York – Tokyo) 1987 p. 211.
[21] Krivoruchko O P, Zhuzhgov A V, Larina T V and Parmon V N 2013 *Dokl. Phys. Chem.* **450** 115

Systematical Analysis of Alpha-active Nuclides

J Munkhsaikhan[1], M Odsuren[1,2] and G Khuukhenkhuu[1]

[1] Nuclear Research Center, National University of Mongolia, Ulaanbaatar, Mongolia
[2] School of Engineering and Applied Sciences, National University of Mongolia, Ulaanbaatar, Mongolia

E-mail: j.munkhsaikhan@num.edu.mn

Abstract. The systematical analysis of alpha-active nuclides is useful for both the nuclear technology applications and understanding of nuclear structure study. Geiger-Nuttall law, which describes a dependence of the disintegration constant on the range of α-particles, was deduced from the Gamow theory explaining the passage of the α-particles through the Coulomb barrier by the quantum mechanical tunneling effect. Ground-to-ground state α-transitions for natural and artificial α-active nuclides were analyzed utilizing the Geiger-Nuttall rule. From rough analysis five group-like branches on the dependence of α-decay half-lives on α-particle energy were observed. Detailed analysis shows that precise linear dependences of the logarithm of α-decay half-lives on the reciprocal of square root of the α-particle energy for even-even isotopes of the U, Pu and Cm there are. However, for some even-even isotopes of the Po, Ra and Th regular behaviour of mass numbers was broken. This non-regularity of the mass numbers on the Geiger-Nuttall line is explained by the nuclear shell model.

1. Introduction

The alpha decay is a disintegration of the radioactive nucleus which emits an α-particle consisting of two protons and two neutrons. The systematical analysis of alpha-active nuclides is useful for both the nuclear technology applications and understanding of nuclear structure study. For example, the alpha decay leads to accumulation of helium gas in the structural materials of nuclear fission and fusion reactors.

In 1911 Geiger and Nuttall established [1] an empirical law which describes a dependence of the disintegration constant on the range of α-particles. Energy of the outgoing α-particle is usually lower than potential energy of the daughter nucleus [2]. Although from the view point of the classical mechanics it is unclear how alpha particle can overcome from the nuclear potential, Gamow theory [3,4] can describe the passage of α-particles through the Coulomb potential barrier by the quantum-mechanical tunneling effect.

In this work the Geiger-Nuttall law deduced from the Gamow theory and ground-to-ground state α-decay data were systematically analyzed using this law and the nuclear shell model deductions.

2. Theoretical background

Geiger and Nuttall law [1] relates the decay constant of a radioactive isotope with the range of the α-particle as following:

$$\log(\lambda) = A \log R_{\alpha} + B. \tag{1}$$

Here: λ is the disintegration constant, $\lambda=0.693/T_{1/2}$, where $T_{1/2}$ is the half-life; R_α is the range of the α-particle, $R_\alpha \sim E_\alpha^n$ where E_α is the α-particle energy; A and B are the constants. Then, the Geiger-Nuttall law can be rewritten as following:

$$\ln(T_{1/2}) = a\frac{1}{\sqrt{E_\alpha}} + b . \quad (2)$$

The formula (2) can be deduced from Gamow theory [3-5]. Penetration probability of α-particle through potential barrier is determined by [5,6]:

$$\ln(T_\alpha) = -2\int_R^{R_0}\sqrt{\frac{2m_\alpha(V(r)-E_\alpha)}{\hbar^2}}dr . \quad (3)$$

Here: m_α is the mass of the α-particle; $V(r)$ is thepotential energy of the daughter nucleus; R and R_o are the inner and outer classical turning points, respectively (Fig.1).

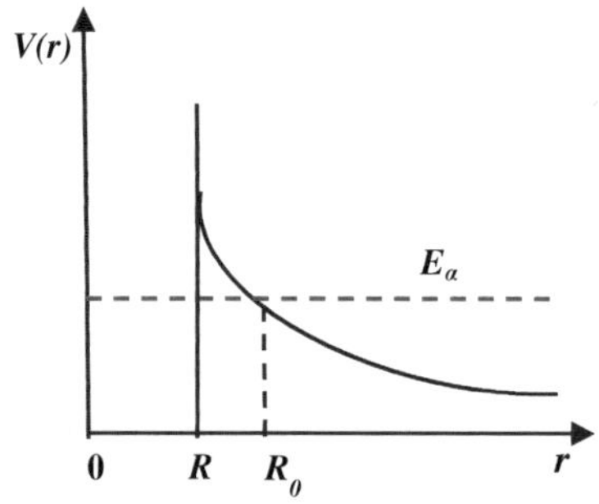

Figure 1. Alpha-particle penetration through the potential barrier.

The inner classical turning point, R, can be obtained for square potential well as a daughter nuclear radius:

$$R = r_0 A_D^{1/3} , \quad (4)$$

where: A_D is the mass number of the daughter nucleus and $r_0=1.25\cdot10^{-13}cm$. For the potential energy of the daughter nucleus, $V(r)$, we can use the Coulomb potential as a first approximation:

$$V(r) = \frac{2Ze^2}{r} , \quad (5)$$

where: e is the elementary charge; and Z is the proton number of the daughter nucleus. Then, from Fig.1 the outer classical turning point can be determined from following expression:

$$\frac{2Ze^2}{R_0} = E_\alpha . \quad (6)$$

So, the formula (3) can be rewritten in the form

$$\ln(T_\alpha) = -2\int_R^{R_0}\sqrt{\frac{2m_\alpha\left[\frac{2Ze^2}{r} - E_\alpha\right]}{\hbar^2}}dr . \quad (7)$$

The following simple substitutions are used to calculate the integral in Eq.(7):

$$x = \frac{r}{R_0} \text{ and } x_0 = \frac{R}{R_0}.(8) \qquad (8)$$

Then, from the expression (7) can be gotten following formula:

$$\ln(T_\alpha) = -\frac{4e^2 Z}{\hbar}\sqrt{\frac{2m_\alpha}{E}}\int_{x_0}^{1}\sqrt{\frac{1}{x}-1}dx \qquad (9)$$

The integral in Eq.(9) can be taken as follows:

$$\int_{x_0}^{1}\sqrt{\frac{1}{x}-1}dx = \int_{0}^{1}\sqrt{\frac{1}{x}-1}dx - \int_{0}^{x_0}\sqrt{\frac{1}{x}-1}dx \approx \int_{0}^{1}\sqrt{\frac{1}{x}-1}dx - \int_{0}^{x_0}\sqrt{\frac{1}{x}}dx - \int_{0}^{1}\sqrt{\frac{1}{x}-1}dx \quad 2\sqrt{x_0} \qquad (10)$$

Here the approximation of x_0<<1 was used. If we use the substitution $x=sin^2\theta$ the integral in Eq.(10) is taken as following:

$$\int_{0}^{1}\sqrt{\frac{1}{x}-1}dx = 2\int_{0}^{\pi/2}\cos^2\theta d\theta = \frac{\pi}{2}. \qquad (11)$$

So, the integral in (9) is given by

$$\int_{x_0}^{1}\sqrt{\frac{1}{x}-1}dx \approx \frac{\pi}{2} - 2\sqrt{\frac{R}{R_0}}. \qquad (12)$$

Then, from Eqs.(7), (9) and (12) the following formula can be obtained

$$\ln(T_\alpha) \approx -\frac{2\pi e^2 Z}{\hbar}\sqrt{\frac{2m_\alpha}{E_\alpha}} + \frac{8}{\hbar}\sqrt{e^2 ZRm_\alpha}. \qquad (13)$$

Taking into account an α-clustering effect, the disintegration constant for α-decay can be expressed as following:

$$\lambda = \phi_\alpha f_\alpha T_\alpha, \qquad (14)$$

where: ϕ_α is the α-clustering factor; f_α is the collision frequency of the α-particle in the potential barrier of the daughter nucleus. In the case of one body approximation [7] the α-clustering factor can be assumed as $f_\alpha=1$. Then, the Eq.(14) can be rewritten in the form

$$\lambda = \frac{0.693}{T_{1/2}} = f_\alpha T_\alpha. \qquad (15)$$

The collision frequency of the α-particle can be obtained as

$$f_\alpha = \frac{\upsilon_\alpha}{2R} = \frac{\sqrt{2E_\alpha / m_\alpha}}{2R}. \qquad (16)$$

From Eqs.(15) and (16) the half-life is given by

$$T_{1/2} = \frac{2R \cdot 0.693}{\sqrt{2E_\alpha / m_\alpha}} \cdot \frac{1}{T_\alpha}. \qquad (17)$$

So, from Eqs.(13) and (17) can be got following expression

$$\ln(T_{1/2}) = \ln\frac{2R\cdot 0.693}{\sqrt{2E_\alpha/m_\alpha}} - \frac{8}{\hbar}\sqrt{e^2 ZRm_\alpha} + \frac{2\pi e^2 Z}{\hbar}\sqrt{\frac{2m_\alpha}{E_\alpha}}. \quad (18)$$

Then, the Eq.(18) can be rewritten in the following form

$$\ln(T_{1/2}) = a\frac{1}{\sqrt{E_\alpha}} + b, \quad (19)$$

where:

$$a = \frac{2\pi e^2 Z}{\hbar}\sqrt{2m_\alpha} \quad (20)$$

and

$$b = \ln\frac{2R\cdot 0.693}{\sqrt{2E_\alpha/m_\alpha}} - \frac{8}{\hbar}\sqrt{e^2 ZRm_\alpha}. \quad (21)$$

It can be seen that the formula (19) is the same as the expression (2) which was directly written from the Geiger and Nuttall law (1). It should be noted that α-particle energy under logarithm is included in the parameter b which can be considered almost constant in comparison with $1/\sqrt{E_\alpha}$ in the Eq.(19). Also, the proton number Z in Eqs.(20) and (21) can be taken as an effective and average value for all considered nuclides. Thus, the Eq.(19) will be utilized for systematical analysis of known experimental data of the α-decay.

3. Result of analysis and discussion

Decay data of the ground-to-ground state α-transitions for over 450 natural and artificial alpha-active nuclides [8-10] including rare-earth and super-heavy elements were analyzed using the Geiger-Nuttall law (19). The dependence of the logarithm of α-decay half-lives, $T_{1/2}$ (sec),on the reciprocal of square root of the α-particle energy, E_α(MeV), for studied isotopes is shown in Fig.2.

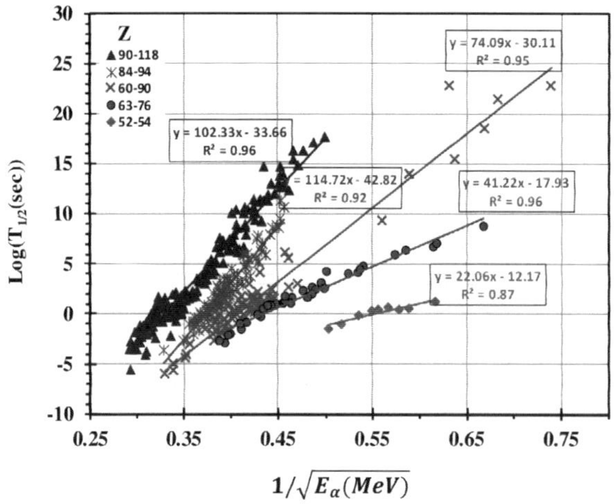

Figure 2. The logarithm of α-decay half-lives versus the reciprocal of square root of the α-particle energy

From the preliminary and rough analysis it was seen that five group-like branches in the dependence of half-life on the α-particle energy were observed [11]. The detailed analysis shows that precise linear dependences of the logarithm of α-decay half-lives on the reciprocal of square root of the α-particle energy for even-even isotopes of the U, Pu and Cm there are (Fig.3). Also, mass numbers of these isotopes are regularly increased along the line corresponding to the Geiger Nuttall law. At the same time for some even-even isotopes of the Po, Ra and Th such regular behaviour of the dependence of the $\ln T_{1/2}$ versus $1/\sqrt{E_\alpha}$ was broken (Fig.4).

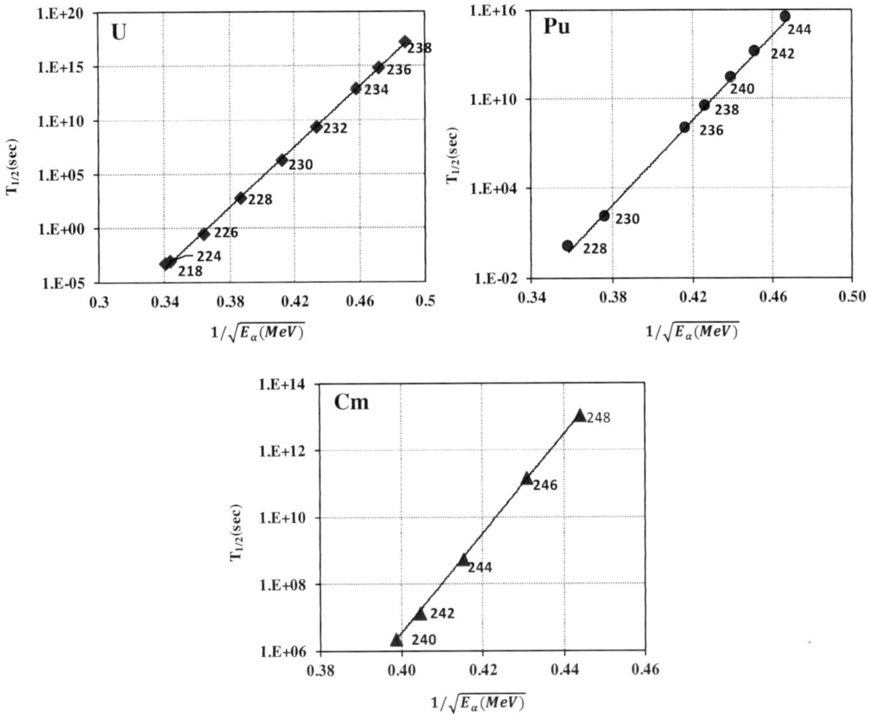

Figure 3. The same as in Fig. 2 for isotopes of the U, Pu and Cm.

It can be seen from Fig.4 that ^{196,198,208,210}Po, ^{214}Ra and ^{216}Th are off the regular behaviour of mass numbers which are increased along the Geiger-Nuttall law line. REN Zhong-Zhou et al. [12] attempted to explain this effect by the nuclear shell model. For the isotopes of ^{210}Po, ^{214}Ra and ^{216}Th number of neutrons is, really, N=126 (magic number) and neutron shell is closed (see Fig.5).

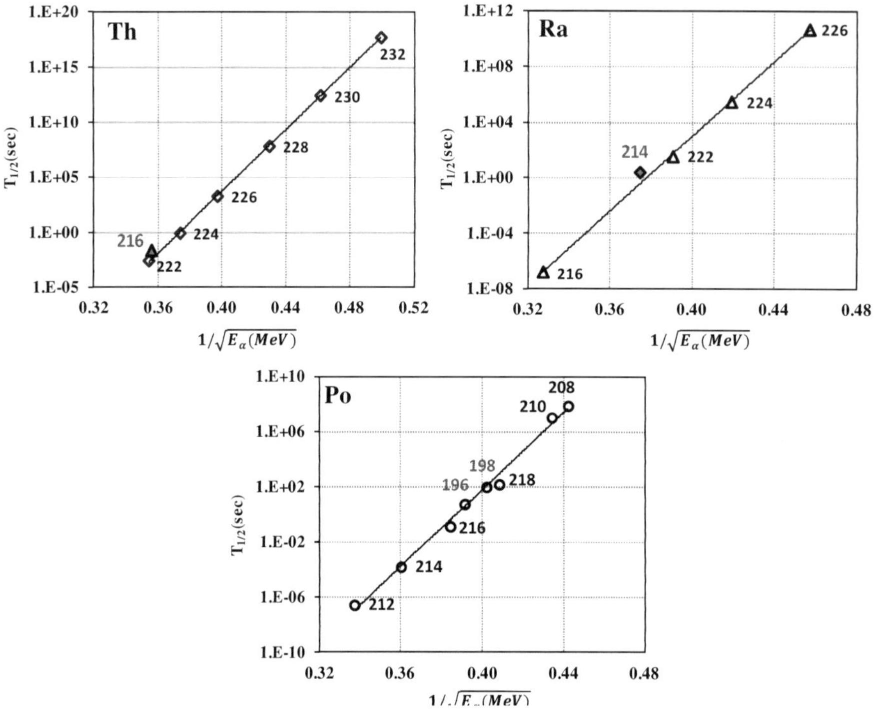

Figure 4. The same as in Fig. 2 for isotopes of the Th, Ra and Po.

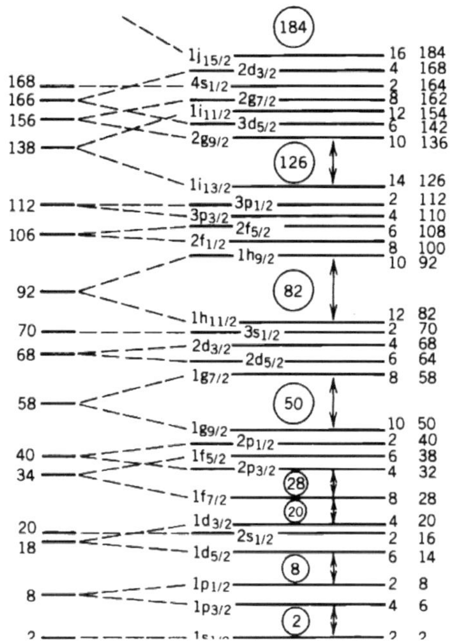

Figure 5. Energy levels of the nuclear shell model [13].

Next energy levels are usually split from the closed shell by appreciable energy gap. So, sudden breaks of the regular behaviour of mass numbers for the ^{208,210}Po, ^{214}Ra and ^{216}Th are, perhaps, caused by the gap of energy levels around magic number N=126. Similar non-regularity for the ^{196,198}Po can be explained by closing the subshell $3P_{1/2}$ (N=112). Also, steep leaps for isotopes ^{252}Cf and ^{254}Fm on the Geiger-Nuttall lines were observed (see Fig.6).

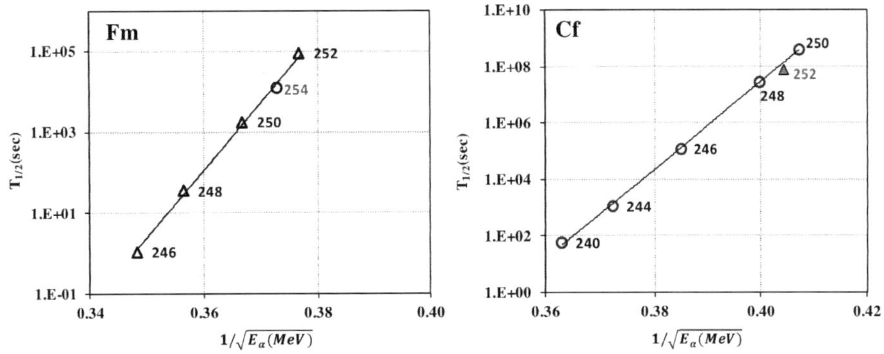

Figure 6. The same as in Fig. 2 for isotopes of the Fm and Cf.

In these cases, the neutron number N=154 and the subshell $1i_{11/2}$ is closed. In addition, a straight relation between the $\ln T_{1/2}$ and $1/\sqrt{E_\alpha}$ appears for isotonic chains with N=124, 126, 150 and 152 [12] (see Fig.7). However, theoretical explanation of this regularity, as far as we know, is not available.

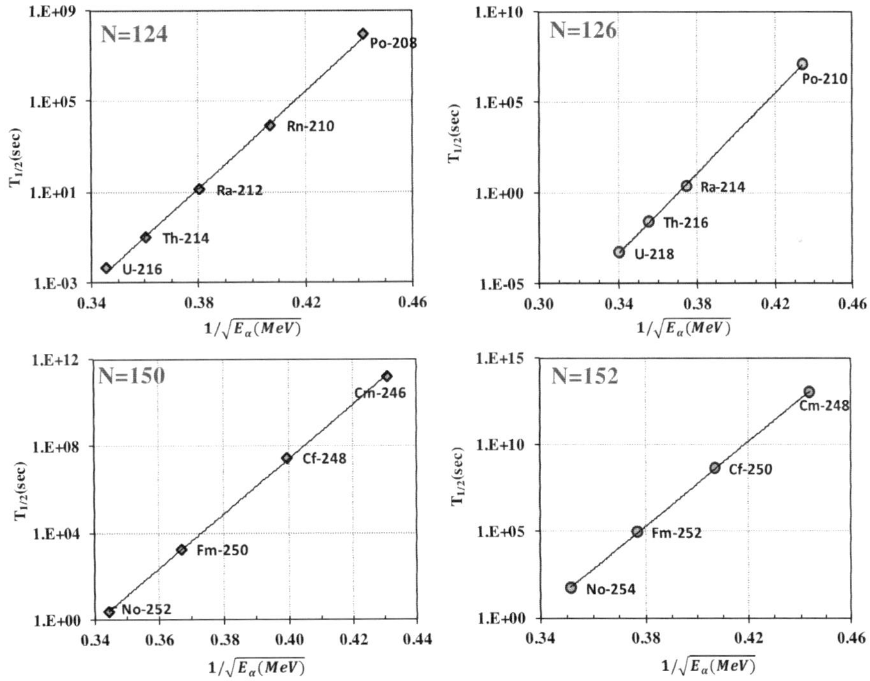

Figure 7. The same as in Fig. 2 for isotonic chains of N=124, 126, 150 and 152.

4. Conclusions

1. The Geiger-Nuttal law was deduced from the quantum Gamow theory. The ground-to-ground state α-transitions for natural and artificial ~450 α-active nuclides were analyzed using the Geiger-Nuttall rule. Five group-like branches for considered nuclides were observed.
2. Precise linear dependence of the logarithm of α-decay half-lives on the reciprocal of square root of the α-particle energy for even-even isotopes of the U, Pu and Cm were established. Mass numbers of the isotopes are regularly increased along the Geiger-Nuttall line.

3. For some even-even isotopes of the Po, Ra and Th regular increasing the mass number along the Geiger-Nuttall line was broken. These results were explained by the nuclear shell model.
4. A straight relation between the $\ln T_{1/2}$ and $1/\sqrt{E_\alpha}$ appears for isotonic chains with N=124, 126, 150 and 152. A theoretical substantiation of this regularity, as far as we know, is not available. So, it is interesting to investigate this effect in the future.

References

[1] Geiger H and Nuttall J M 1911 *Phil. Mag.*Series 6**22**p 613.
[2] Perlman I and Rasmussen J O1957 *Alpha Radioactivity* (Berlin:Springer-Verlag)
[3] Gamow G1928 *Z.Phys.,* No.3-4, **51** p 204
[4] Gamow G 1928 *Nature,* Issue 3082, **122**p 805
[5] Wichmann E H 1971 *Quantum Physics, Berkeley Physics Course,*(New York:McGraw-Hill Book Company)**4**
[6] Rasmussen J O 1959 *Phys. Rev.,* No.6, **113**p 1593
[7] HannaG C 1959 *Alpha-Radioactivity. In book: Experimental Nuclear Physics,* ed SergeE(New York-London)**3**
[8] FirestoneR B, Frank S Y, Chu and Baglin C M 1998 *Table of Isotopes*, 8th Edition (New York: John Wiley and Sons)
[9] Baum E M*et al.*2010 *Chart of Nuclides*, 17th Edition (Knolls Atomic Power Laboratory)
[10] Tuli J K 2011*Nuclear Wallet Cards*(New York: BNL)
[11] Delgersaikhan T,Munkhsaikhan J, KhuukhenkhuuG and Turbold A2013*Proc. 5th Int. Conf. on Contemporary Physics (*Ulaanbaatar) p 238
[12] Ren Zhong-Zhou, Tai Fei and Shen Wen-Qing2003 *Commun.Theor.Phys.***40**p 191
[13] Krane K S1987*Introductory Nuclear Physics*(USA: Publisher Wiley, University of Michigan)

Acknowledgement
This work is supported by the Mongolian Science and Technology Foundation: contract No ShuSs-2019/06.

IOP Conf. Series: Materials Science and Engineering 1000 (2020) 012003 doi:10.1088/1757-899X/1000/1/012003

A new approach to the criterion of the liquid-glass transition

D S Sanditov[1,2] *, S Sh Sangadiev[1], M V Darmaev[1]

[1] Buryat State University, 67000 Ulan-Ude, Smolin st. 24a, Russia;
[2] Institute of Physical Materials Science SB RAS,
670031, Ulan-Ude, Sakhyanovoy st., 6, Russia.
*e-mail: sanditov@bsu.ru

Annotation. A liquid transforms into a frozen glassy state when its cooling rate q = (dT/dt) reaches a certain fraction of C_g of the characteristic cooling rate $q_g = (T_g/\tau_g)$, which is closely related to the structure relaxation time τ_g at the glass transition temperature T_g.
Key words: generalized criterion, glass transition temperature, relaxation time, cooling rate, model, atomic delocalization.

Elucidation of the nature of the glass transition of liquids is the most difficult problem in condensed matter physics (see, for example, [1-4]).

This work is devoted to the analysis and generalization of kinetic criteria for glass transition using the model of delocalized atoms [5, 6]. The characteristic cooling rate of the glass-forming melt is introduced and a generalized criterion for the liquid-glass transition is proposed.

From the point of view of the relaxation approach, in the process of glass transition of a liquid, the decisive role is played by the relationship between the relaxation time of the structure τ and the cooling rate of the melt q. In 1951, Bartenev [7], proceeding from general considerations, proposed the following relationship between these quantities (the glass transition equation)

$$q\tau_g = C, \tag{1}$$

where τg is the relaxation time at the glass transition temperature T_g, C – is an empirical parameter. Here and below, q should be understood as the absolute value of the rate of temperature change |q| when cooling the melt or when heating the glass.

Volkenshtein and Ptitsyn [8] developed a physical theory according to which particles of a glass-forming liquid can be in two states separated by an energy barrier. The solution of the constructed kinetic equation leads to the conclusion that the criterion for the liquid-glass transition is the following equation

$$q\tau_g = \delta T_g. \tag{2}$$

The δT_g value is interpreted as a temperature band characterizing the interval of the transition from liquid to glass upon cooling.

Schmelzer et al. [1, 9] introduce the characteristic time of temperature variation $\tau_T = T/q$ and consider the relationship between τ_T and the relaxation time of the structure τ. According to Schmelzer's assumption, the liquid passes into a glassy state when, upon cooling, the time of structural relaxation τ increases and becomes equal to the characteristic time of temperature change $\tau \approx \tau_T$. Proceeding from these considerations, taking into account q = (dT/dt), Schmelzer formulated the following kinetic criterion for glass transition [1, 9].

$$\frac{\tau}{\tau_T} \approx 1, \quad \left[\frac{1}{T}\left(\frac{dT}{dt}\right)\tau\right]_{T=T_g} = C_3, \quad C_3 \approx 1. \tag{3}$$

Let us dwell on the analysis of the indicated glass transition criteria using the model of delocalized atoms [5, 6].

The temperature dependence of the structural relaxation time in the glass transition region $\tau(T)$ is successfully described by the Williams-Landel-Ferry (WLF) equation [10]

$$\ln\frac{\tau(T)}{\tau(T_g)} = -C_1\frac{T - T_g}{T - T_g + C_2}.$$

Substitution of $\tau(T)$ from this equation into the Volkenstein – Ptitsyn relation [8]

$$\left.\frac{d\tau}{dT}\right|_{T=T_g} = -\frac{1}{q} \qquad (4)$$

leads to equality

$$q\tau_g = \frac{C_2}{C_1}, \qquad (5)$$

where the parameters of the WLF equation C_1 and C_2 within the framework of the model of delocalized atoms have the following physical meaning [5]

$$C_1 = \frac{1}{f_g}, \qquad C_2 = \frac{f_g}{\beta_f}, \qquad (6)$$

moreover, the product of T_g and the coefficient of thermal expansion of the fluctuation volume at the glass transition temperature β_f is a single-valued function of the fraction of the fluctuation volume f_g frozen at the glass transition temperature [5],

$$\beta_f T_g = f_g \ln(1/f_g). \qquad (7)$$

From the above expressions (5) - (7) we obtain the following interpretation of the glass transition equation

$$q\tau_g = \frac{f_g}{\ln(1/f_g)}T_g. \qquad (8)$$

The fluctuation volume of an amorphous substance ΔV_e is due to thermal displacements of atoms [5]

$$\Delta V_e = N_e \Delta v_e,$$

where N_e is the number of delocalized atoms (kinetic units), Δv_e is the elementary fluctuation volume required for the delocalization of an atom - its limiting displacement from the local equilibrium position. The volume fraction of the fluctuation volume f_g frozen at the glass transition temperature T_g weakly depends on the nature of amorphous substances [5, 6] (Table 1)

$$f_g = \left(\frac{\Delta V_e}{V}\right)_{T=T_g} \approx 0.020 - 0.030.$$

For glasses of one class, the value of f_g turns out to be an almost universal constant $f_g \approx$ const. The constancy of f_g is valid with the accuracy of the constancy of the logarithm of the ratio of the limiting values of viscosity [11]: $1/f_g = \ln(\eta_g/\eta_0)$, where $\eta_0 \approx$ const is the high-temperature limit of viscosity, $\eta_g \approx$ const is the viscosity at the glass transition temperature (the rule of constancy of viscosity at $T = T_g$).

Dividing both sides of equality (8) by the glass transition temperature T_g, we arrive at the generalized kinetic glass transition criterion

$$\frac{q\tau_g}{T_g} = C_g, \qquad C_g = \frac{f_g}{\ln(1/f_g)} \approx 7\cdot 10^{-3}, \qquad (9)$$

which can be written as an expression for determining the transition temperature T_g

$$\left.\left(\frac{1}{T}\left|\frac{dT}{dt}\right|\tau\right)\right|_{T=T_g} = C_g, \quad C_g \approx 7\cdot 10^{-3}. \qquad (10)$$

When estimating C_g, we used the averaged value $f_g \approx 0.025$ (Table 1).

Relation (10) can be considered as a justification and generalization of the Schmelzer glass transition criterion (3). In this case, the constant C_3 acquires a certain physical meaning

$$C_3 = C_g = \frac{f_g}{\ln(1/f_g)} \approx \text{const} \approx 7\cdot 10^{-3}.$$

This eliminates the drawback of criterion (3), which is reduced to equating the constant C_3 to one: $C_3 \approx 1$, and in fact in the form of a postulate without any special justification. It is easy to see that

Schmelzer's criterion (3) leads to an erroneous equality: $q\tau_g \approx T_g$, which contradicts the glass transition equation (2), which follows from the classical Volkenstein-Ptitsyn theory. For example, for silicate glasses at $T_g \approx 800$ K, the product $q\tau_g$, according to criterion (3), should correspond to the value: $q\tau_g \approx 800$ K, while the typical values of the product $q\tau_g$ for these glasses are about [6]: $q\tau_g \approx (5\div10)$ K.

Table 1

Parameters of the Williams – Landel – Ferry equation C_1, C_2 and glass transition characteristics of amorphous substances (data from [6] were used)

Amorphous substance	T_g, K	C_1	C_2, K	$\delta T_g = \frac{C_2}{C_1}$, K	$f_g = \frac{1}{C_1}$	$C_g \cdot 10^3$
1	2	3	4	5	6	7
Sodium silicate glass Na_2O-SiO_2.						
Na_2O, mol.% 15	782	36	430	12	0.028	7.8
20	759	36	390	11	0.028	7.8
25	739	35	355	10	0.028	7.8
30	721	35	322	9	0.028	7.8
33	712	35	304	9	0.028	7.8
35	705	35	291	8	0.028	7.8
Amorphous polymers						
Polyisobutylene	202	38	104	2.7	0.026	7.1
Polyvinyl acetate	305	36	47	1.3	0.028	7.8
Polyvinylchloroacetate	296	40	40	1.0	0.025	6.8
Polymethyl acrylate	276	42	45	1.1	0.024	6.4
Polyurethane	238	36	33	0.9	0.028	7.8
Natural rubber	300	38	54	1.4	0.026	7.1
Methacrylate polymers						
ethyl	335	40	65	1.6	0.025	6.8
n-butyl	300	39	97	2.5	0.026	7.1
n-butyl	253	37	107	2.9	0.027	7.5
Metallic glass						
$Pd_{40}Ni_{40}P_{20}$	602	39	93	2.4	0.026	7.1
$Pt_{60}Ni_{15}P_{25}$	500	37	95	2.6	0.027	7.5
$Pd_{77.5}Cu_6Si_{16.5}$	653	38	100	2.6	0.026	7.1
$Fe_{80}P_{13}C_7$	736	38	120	3.2	0.026	7.1
Low molecular weight organic glasses						
Propanol	98	41	25	0.6	0.024	6.4
Propylene glycol	160	44	40	0.9	0.023	6.1
Glycerol	185	42	53	1.3	0.024	6.4
Rosin	303	36	29	0.8	0.028	7.8
Isobutyl alcohol	118	38	38	1.0	0.026	7.1

The generalized kinetic criterion for glass transition (10) as applied to silicate glasses for the product qτg gives the values ($T_g \approx 700\div800$ K)

$$q\tau_g = C_g T_g \approx 5 \div 6\ \text{K},$$

which are in agreement with the typical data for the indicated glasses.

The generalized criterion for the liquid glass transition (9) proposed by us can be formulated as follows: a liquid passes into a frozen glassy state when its cooling rate q reaches a certain fraction of C_g of the characteristic cooling rate $q_g = (T_g/\tau_g)$, which is closely related to the relaxation time of the structure τ_g at glass transition temperature T_g,

$$\frac{q}{q_g} = \frac{q}{(T_g/\tau_g)} = C_g. \tag{11}$$

A similar, but slightly different formulation is possible: the liquid glassy when the relaxation time of the structure τ_g reaches a fraction of C_g of the characteristic time of temperature change $\tau_T = (T_g/q)$:

$$\frac{\tau_g}{\tau_T} = \frac{\tau_g}{(T_g/q)} = C_g \,. \tag{12}$$

Thus, a generalization of the kinetic criteria for glass transition of Bartenev (1), Volkenstein - Ptitsyn (2), Schmelzer (3) has been proposed. The generalized kinetic criterion for glass transition (9) includes the relaxation time τ_g, the cooling rate q, the transition temperature T_g, and the almost universal value of C_g, which is determined by the volume fraction of the fluctuation volume f_g frozen at the glass transition temperature T_g.

Acknowledgment
This work was supported by the Ministry of Science and Higher Education of the Russian Federation (Grant No. 3.5406.2017 / 8.9).

References

1. Tropin T V, Schmelzer J W and Aksenov V L, 2016 *Phys.-Usp.* **59**, 42.
2. Sanditov D S and Ojovan M I, 2019 *Phys.-Usp.* **62**, 111.
3. Sanditov D S and Bartenov G M, *Physical Properties of Disordered Structures,* Nauka, Novosibirsk, 1982.
4. Rostiashvili V G, Irzhak V I and Rozenberg B A, *Glass Transitions in Polymers*, Khimiya, Leningrad, 1987.
5. Sanditov D S, 2012 *J. Exp. Theor. Phys.* **115**, 112.
6. Sanditov D S, 2016 *J. Exp. Theor. Phys.* **123**, 429.
7. Bartenev G M, 1951 *Dokl. Akad. Nauk SSSR* **76**, 227.
8. Vol'kenshtein M V and Ptitsyn O B, 1956 *Zh. Tekh. Fiz.* **26**, 2204.
9. Schmelzer J W P, 2012 *J. Chem. Phys* **136**, 074512.
10. Ferry J D, *Viscoelastic Properties of Polymers,* John Wiley & Sons, Inc., New York, 1980.
11. Sanditov D S, Razumovskayac I V and Mashanov A A, 2020 *Polym. Sci., Ser. A* **62**, 588.

On the activation energy of the glass transition process

D S Sanditov [1,2] *, A A Mashanov [1], S S Badmaev [1]

[1] Buryat State University, 67000 Ulan-Ude, Smolin st. 24a, Russia;
[2] Institute of Physical Materials Science SB RAS,
670031, Ulan-Ude, Sakhyanovoy st., 6, Russia.
*e-mail: sanditov@bsu.ru

Annotation. Calculation of the temperature dependence of the activation energy of the glass transition process U(T) of amorphous organic polymers and inorganic glasses with the use of the Williams-Landel-Ferry equation for the relaxation time (viscosity) is proposed. The estimates made are in satisfactory agreement with experiment. The nature of the U(T) dependence is discussed.
Key words: glass transition, viscosity, relaxation time, activation energy.

Introduction

The nature of the transition of an amorphous substance from a liquid to a glassy state remains one of the unsolved problems. The glass transition of a liquid is closely related to an increase in viscosity upon cooling the melt. In this case, the activation energy of the glass transition process, as a rule, coincides with the activation energy of a viscous flow (relaxation time) in the region of the liquid-glass transition.

The temperature dependence of the viscosity of simple low-viscosity liquids such as acetone is successfully described by the equation of Ya.I. Frenkel [1]

$$\eta = \eta_0 \exp\left(\frac{U}{RT}\right) \quad (1)$$

with constant activation energy U = const. Here η_0 is the preexponential factor, R – is the gas constant. However, for glass-forming melts in formula (1), the activation energy in the glass transition region rather strongly depends on temperature: U = U(T). In this regard, empirical viscosity equations are proposed, which implicitly take into account this dependence U(T). Among them, the WLF (Williams-Landel-Ferry) ratio is widespread [2, 3]

$$\ln a_T = -C_1 \frac{T - T_g}{T - T_g + C_2}, \qquad a_T = \frac{\tau(T)}{\tau(T_g)} \cong \frac{\eta(T)}{\eta(T_g)}, \quad (2)$$

where C_1 and C_2 are empirical constants, T_g is the glass transition temperature, τ is the relaxation time. The validity of this equation has been shown in many works for various vitrifying systems [2-4].

Earlier, a conclusion and justification of this expression was proposed without specifying the functions $\tau(T)$ and $\eta(T)$ [5].

The present work is devoted to the assessment of the temperature dependence of the activation energy of the glass transition process U(T) using data on the parameters of the WLF equation C_1 and C_2.

Calculation of the temperature dependence of the activation energy of glass transition using the Williams-Landel-Ferry equation

It is easy to verify that the WLF equation (2) is algebraically derived from the well-known Vogel – Fulcher – Tamman relation [4]

$$\eta = \eta_0 \exp\left(\frac{B_o}{T - T_o}\right), \tag{3}$$

where the empirical constant Bo has the dimension of temperature and is measured in kelvin. Moreover, their parameters turn out to be related by the equalities

$$B_o = C_1 C_2, \tag{4}$$

$$T_o = T_g - C_2. \tag{5}$$

From a comparison of formulas for viscosity (1) and (3), assuming the preexponential factors in them to be equal, we obtain the following dependence of the activation energy of viscous flow on temperature

$$U = \frac{B_o RT}{T - T_o}. \tag{6}$$

Since for many amorphous polymers and inorganic glasses the values of the parameters of the WLF equation C_1 and C_2 are known [2 - 4], using expressions (4) and (5), it is expedient to pass from dependence (6) to the value U(T) as a function of these parameters

$$U = \frac{C_1 C_2 RT}{T - T_g + C_2}. \tag{7}$$

This equality implies a simple relation for calculating the activation energy of the glass transition process $U_g = U(T_g)$ at the glass transition temperature $T = T_g$

$$U_g = C_1 R T_g. \tag{8}$$

For sheet silicate glass, for example, C_1 and T_g are equal, respectively [6]: $C_1 = 36$ and $T_g = 807$ K, whence we have ($R = 8.3$ J/(mol·K)): $U_g = 241$ kJ/mol. The methods of relaxation spectrometry [4, 6] for this glass lead to practically the same value $U_g = 242$ kJ/mol.

It is known that many vitrifying systems pass from a liquid to a solid glassy state at practically the same viscosity $\eta_g \approx$ const $\approx$ 1012 Pa·s (an approximate rule of constancy of viscosity at T_g). The high-temperature viscosity limit $\eta_0 \approx \eta\ (T \rightarrow \infty)$ is also in the first approximation constant $\eta_0 \approx$ const $\approx 10^{-3.5}$ Pa·s. Therefore, the activation energy at the glass transition temperature $U_g = U(T_g)$ can be estimated by formula (1) at $T = T_g$, $\eta_g \approx \eta\ (T_g) \approx$ 1012 Pa·s and $\eta_0 \approx 10^{-3.5}$ Pa·s [4]

$$U_g = \left(\ln \frac{\eta_g}{\eta_o}\right) RT_g \approx 36 RT_g, \tag{9}$$

which practically coincides with relation (8).

Consequently, the weak dependence of the parameter of the WLF equation C_1 on the nature of the glasses: $C_1 \approx$ const $\approx$ 35 - 40 (Table) [2, 4] is explained by the approximate constancy of the viscosities η_g and η_0:

$$C_1 = \ln\left(\frac{\eta_g}{\eta_0}\right).$$

Table

Parameters of the WLF equation (2) and the activation energy of the glass transition process at $T = T_g$ for amorphous organic polymers and inorganic glasses (data from [2, 4] were used)

N	Amorphous substances	T_g, K	C_1	C_2, K	U_g kJ/mol
1.	$PbO - SiO_2$ PbO, mol. % 42.07	711	36	250	212
2.	$Na_2O - B_2O_3$ Na_2O, mol. % 2.8	544	39	130	131
3.	Sheet glass	807	36	305	241
4.	Polyisobutylene	202	38	104	64
5.	Polyvinyl acetate	305	38	104	96
6.	Polyurethane	238	36	33	71
7.	Natural rubber	300	38	54	95

Note: U_g - calculation by formula (8).

Thus, to calculate U(T) in the glass transition region using formula (7), it is necessary to know the values of three quantities: C_1, C_2, and T_g. Data on the parameters of the WLF equation C_1 and C_2, as well as on the glass transition temperature T_g, for many amorphous organic polymers are available in the book by Ferry [2], and for inorganic glasses - in the monograph by Sanditov and Bartenev [4]. One of the ways to define C_1 and C_2 is described in this book [4, p. 86].

In fig. 1 shows the temperature dependence of the activation energy of the glass transition process for sheet silicate glass. Curve - calculation by formula (7), and points - experimental data. The experimental points were obtained from the data on the viscosity logη(T) using Eq. (1), from which the dependence U(T)

$$U = 2.3RT\left[\lg\eta g\eta(- \lg\eta_0\right] \tag{10}$$

As you can see, the points fall on the calculated curve, which confirms the applicability of the WLF formula. Similar U – T plots were constructed by us for a number of organic amorphous polymers (see, for example, for polyvinyl acetate, Fig. 2).

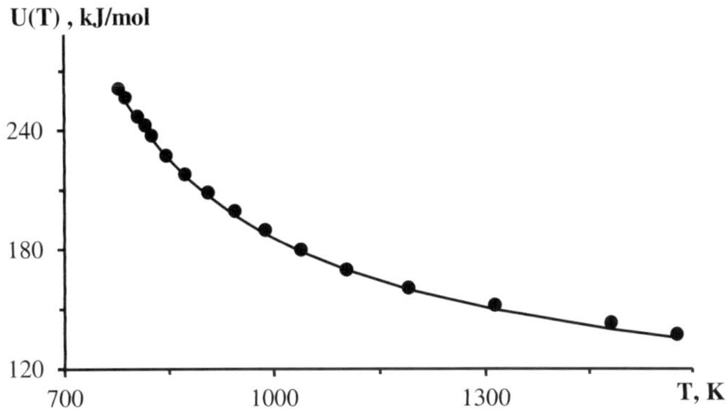

Figure 1. Temperature dependence of the activation energy of glass transition U(T) for sheet silicate glass. Points - experimental data, curve - calculation by formula (7).

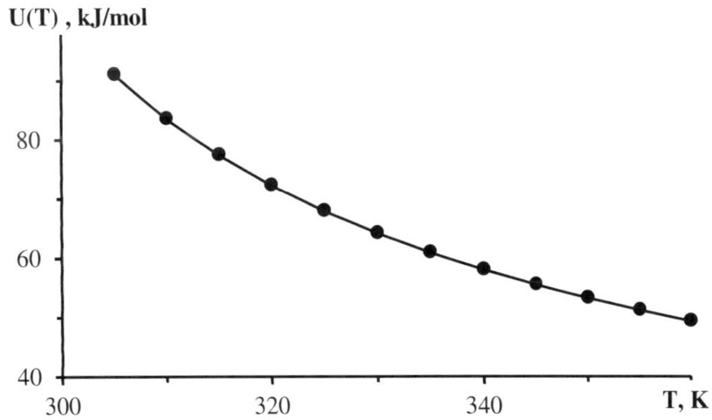

Figure 2. Temperature dependence of the activation energy of glass transition U(T) for polyvinyl acetate. Points - experimental data, curve - calculation according to equation (7).

On the nature of the U(T) dependence

The main regularities of the glass transition process are qualitatively the same for the vast majority of amorphous substances, regardless of their nature: for inorganic glasses, amorphous organic polymers, metal amorphous alloys, aqueous solutions, chalcogenides, which is confirmed by the presence of universal equations and rules in the glass transition region [4-7].

We proceed from the idea that the physical reason for the sharp increase in the activation energy of the glass transition process with decreasing temperature is the configurational change in the structure of the glass-forming melt.

Within the framework of the model of delocalized atoms [8, 9] for glass-forming liquids, the activation energy of the glass transition (viscous flow) in Eq. (1) is the sum of two terms

$$U = U_\infty + U_s(T), \quad (11)$$

where U_∞ is the potential for the transition of a kinetic unit (atom, molecule) to a new local position, $U_S(T)$ is the potential for a configurational structural change, which is a function of temperature,

$$U_s(T) = RT\left[\exp\left(\frac{\Delta\varepsilon_e}{RT}\right) - 1\right]. \quad (12)$$

Here, $\Delta\varepsilon_e$ is the delocalization energy of the atom (its limiting displacement from the local equilibrium position). The delocalization of an atom is caused by the rearrangement of neighboring particles (fluctuation of short-range order) and reflects a local configurational change in the structure.

At elevated temperatures RT >> $\Delta\varepsilon_e$, the potential of the configurational structural change is equal to zero $U_S(T) = 0$, and the viscosity equation (1), taking into account (11) and (12), transforms into the usual Arrhenius (Frenkel) dependence with a constant activation energy $U = U_\infty$

$$\eta = \eta_0 \exp\left(\frac{U_\infty}{RT}\right). \quad (13)$$

From this point of view, the value of U_∞ turns out to be the high-temperature limit of the activation energy for viscous flow.

At low temperatures, in the glass transition region, the energy of atomic delocalization $\Delta\varepsilon_e$ becomes comparable to the energy of thermal vibrations of the lattice (~ 3RT) and the relative number of delocalized atoms (N_e/N) responsible for fluidity above T_g significantly decreases according to the law [9]. The structure of the melt becomes denser, and for the activation jump of the atom to a new position, a preliminary local configurational change ("loosening") of the structure near T_g is required: the potential for configurational structural change $U_S(T)$ sharply increases. This explains the almost exponential growth of the activation energy of fluidity in the glass transition region.

Thus, the configurational change in the structure, which is described within the framework of the model of delocalized atoms, turns out to be responsible for the temperature dependence of the activation energy of the glass transition process (viscous flow) in the region of the transition from the liquid state to the glassy state.

Conclusion

For amorphous organic polymers and inorganic glasses, methods of calculating the temperature dependence of the activation energy of the glass transition process U(T) are considered. This dependence U(T) is satisfactorily described near T_g using the well-known Williams-Landel-Ferry equation for viscosity (relaxation time). Within the framework of the model of delocalized atoms, a sharp increase in the activation energy in the glass transition region is explained by an increase in the potential for a configurational change in the structure upon cooling the glass-forming melt.

Acknowledgment

This work was supported by the Ministry of Science and Higher Education of the Russian Federation (Grant No. 3.5406.2017 / 8.9).

References

1. Frenkel' Ya I, *Kinetic Theory of Liquids*, Akad. Nauk SSSR, Moscow, 1945.
2. Ferry J D, *Viscoelastic Properties of Polymers,* John Wiley & Sons, Inc., New York, 1980.
3. Williams M L, Landel R F and Ferry J D, 1955 *J. Amer. Chem. Soc.* **77**, 3701.
4. Sanditov D S and Bartenov G M, *Physical Properties of Disordered Structures,* Nauka, Novosibirsk, 1982.
5. Sanditov D S and Razumovskaya I V, 2018 *Polym. Sci., Ser. A* **60**, 156.
6. Bartenev G M and Sanditov D S, *Relaxation Processes in Glassy Systems,* Nauka, Novosibirsk, 1986.
7. Rostiashvili V G, Irzhak V I and Rozenberg B A, *Glass Transitions in Polymers*, Khimiya, Leningrad 1987.
8. Sanditov D S, 2012 *J. Exp. Theor. Phys.* **115**, 112
9. Sanditov D S and Ojovan M I, 2019 *Phys.-Usp.* **62**, 111

IOP Conf. Series: Materials Science and Engineering 1000 (2020) 012005 doi:10.1088/1757-899X/1000/1/012005

Dynamic viscosity of dispersion of silica dioxide nanoparticles

T S Dembelova, S A Balzhinov, D N Makarova, Ye D Vershinina, S B Bazarova and B B Badmaev

Institute of Physical Materials Science, Siberian Branch of the Russian Academy of Sciences, 670047, Ulan-Ude, Russia

E-mail: tu_dembel@mail.ru

Abstract. Dynamic viscosity of silica dioxide suspensions in oils was investigated by acoustical, rotary and capillary methods. An installation has also been created for measuring the viscosity of a liquid at low gradients of the flow velocity, when one can expect that the structure of the liquid change little during elementary acts of viscous flow. It is shown that at low flow rate gradients of the investigated suspensions, an increased value of viscosity is found (relative to the viscometric one), which indicates the structuring of these suspensions with the formation of a supramolecular structure. The observed phenomenon of increased viscosity can be of practical importance for all processes where slow flows prevail.

1. Introduction

Nanoscale objects significantly change the physical properties of both planar and bulk solid and liquid nanostructured systems. Nanoparticles as a dispersed phase lead to the appearance of a volumetric spatial structural network throughout the entire volume of the dispersed system. In this case, the structural and mechanical properties of the dispersed system change, and the relaxation processes occurring in them become more complicated.

2. Materials

Polyorganosiloxane PES-2 and castor oils are used as base liquids. The suspensions were prepared using silica dioxide nanopowder SiO_2 of Tarkosil trademark obtained by gas-phase synthesis [1]. Depending on the peculiarities of the method of obtaining, nanoparticles characterized by the presence on the surface of OH-groups with a concentration of the order of $(0.5 \div 4)$ / nm^2 [2]. The properties of dispersed systems depend on the degree of dispersion of nanoinclusions. Nanoparticles have high surface energy and form agglomerates in a dispersed medium to several microns in size. Figure 1 shows SEM images of the powders obtained using a JSM-6510LV JEOL scanning electron microscope. It can be seen that with an increase in the specific surface of the particles, the sizes of the agglomerates grow, much exceeding the primary sizes of the particles of silicon dioxide. To obtain a stable suspension, it is important to thoroughly disperse it, which allows the nanoparticle agglomerates to be destroyed as much as possible. In addition, during dispersing, a uniform distribution of nanoparticles in the volume is established. Our suspensions were prepared by the ultrasonic method, the optimal dispersing time of 30 min for SiO_2/PES-2 has been determined experimentally. As shown by the Nano Particle Size Analyzer SALD-7500nano the average size of particles in the suspension is 78 nm.

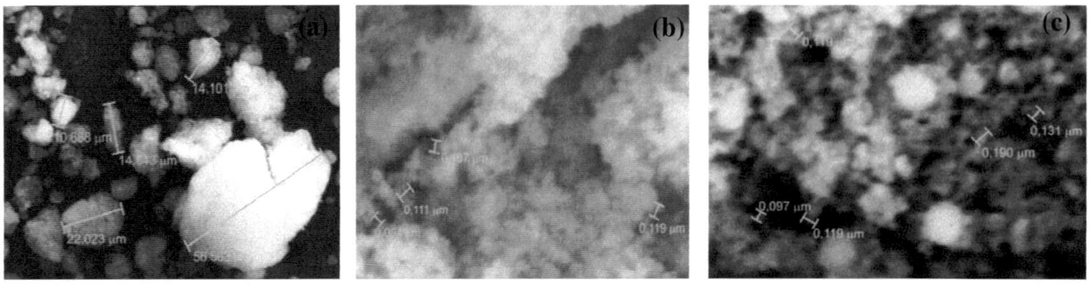

Figure 1. Electronic photograph of silica dioxide nanoparticles SiO_2:
(a) T-20 (20 nm); **(b)** T-05 (50 nm); **(c)** SiO_2 (100 nm).

3. Measurement of dynamic viscosity by rotary and capillary methods

The study of the dynamic viscosity coefficient of suspension of T-0.5 silica dioxide particles in PES-2 liquid was carried out on an Anton Paar MCR 52 rheometer with the use of a cone-plane measuring system with a gap of 0.1 mm. The cone diameter was 50 mm, the angle between the generatrix of the cone and the measurement plane was 1°. The measurements were carried out at a temperature of 23 °C. A linear dependence of the shear stress P on the rate of shear strain $\dot{\gamma}$ was obtained, indicating the Newtonian behavior of the viscosity of these suspensions (figure 2).

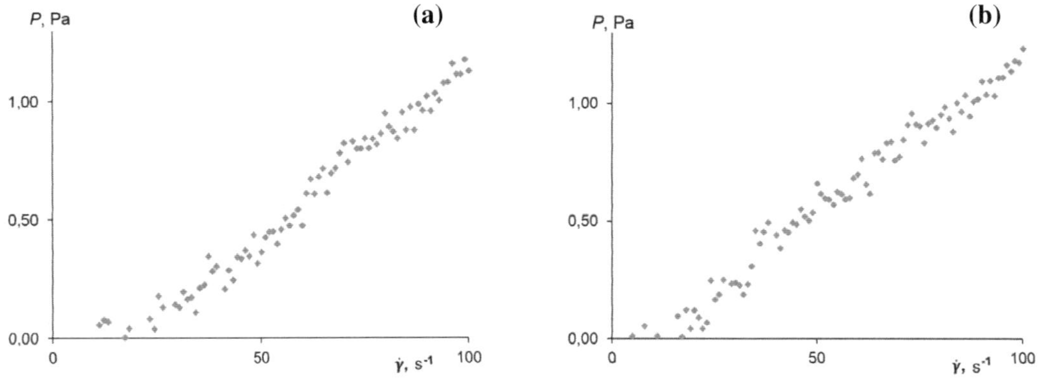

Figure 2. Dependence of the shear stress on the shear rate for the suspensions:
(a) SiO_2/PES-2, 1 wt.%; **(b)** SiO_2/PES-2, 1.25 wt.%.

The viscosity coefficients of the studied liquids were measured with a VPZh-2 capillary viscometer at a constant temperature of 23 °C. Table 1 shows the values of the suspension viscosity coefficients obtained using Anton Paar MCR 52 rheometer and VPZh-2 viscometer. It can be seen that the results of measuring the viscosity of suspensions by these methods are in good agreement with each other within the measurement accuracy. It can also be noted that the viscosity of 1.25 wt.% suspension is slightly higher than the viscosity of the 1.0 wt.%. Numerous works [3, 4] have shown that the viscosity of the suspension increases with an increase in the concentration of nanoparticles.

Table 1. The values of the viscosity coefficients of suspensions SiO_2/PES-2.

Viscometer	t, °C	η, mPa·s SiO_2/PES-2, 1.25 wt.%	SiO_2/PES-2, 1.0 wt.%
Anton Paar MCR 52	23	12.92	11.48
VPZh-2	23	11.74	11.82

4. Investigation of the viscoelastic properties of suspensions by acoustic resonance method

Low-frequency viscoelastic properties of suspension of silica dioxide nanoparticles in PES-2 are studied by the acoustic resonance method using a piezoelectric quartz resonator. Piezoquartz crystal makes oscillations on the main resonance frequency, 73.2 kHz. One end of the piezoquartz horizontal surface contacts with liquid layer covered by quartz cover-plate. The liquid layer tests only shear deformations and the stagnant shear waves are installed in it. Depending on a thickness of the layer, the parameters of the resonance curve of the piezoquartz vary. The theory of the method gives the following expression for complex shift of the resonant frequency [5-7]:

$$\Delta f^* = \frac{Sk^*G^*}{4\pi^2 Mf_0} \cdot \frac{1+\cos(2k^*H - \varphi^*)}{\sin(2k^*H \quad \varphi^*)}, \qquad (1)$$

where $G^* = G' + iG''$ – is the complex shear modulus of the liquid, S – is the cover-plate base area, H – is the thickness of liquid interlayer, M – is the resonator mass, f_0 – is its resonance frequency, k^* – is the complex wave number; φ^* – complex phase shift in the reflection of a shear wave from the liquid – cover-plate interface.

Studies of the low-frequency shear elasticity of colloidal suspensions of nanoparticles have shown that a region of linear elasticity is observed at low oscillation amplitudes. The ratio of the amplitude of the piezoquartz oscillation to the thickness of the liquid layer can be a measure of the angular deformation for small angles. At small deformation angles, the components of the complex shear modulus are constant, then, with an increase in the shear angle, the real modulus decreases, and imaginary one passes through a maximum. It can be assumed that the liquid has an equilibrium supramolecular structure and, at small angles of deformation equilibrium structure remains unchanged, and a region of linear elasticity is found [8]. At a critical shear angle, which corresponds to a certain critical shear stress, the equilibrium structure of suspension begins to break and its viscoelastic properties change. Other liquids, such as glycols, alcohols, acids, exhibit similar behavior [9-11].

Figure 3 presents dependences of the effective viscosity calculated using Maxwell's rheological model on the shear stress for suspensions SiO_2/PES-2 of 1.25 wt.%. At small stresses not exceeding the critical value, a constant increased viscosity characterizes effective viscosity and its value is higher for smaller particles size in suspension. When the structure is destroyed with shear stress increasing, the viscosity decreases to the smallest value. Therefore, it is of great interest to directly measure the increased viscosity of suspensions of nanoparticles with an intact equilibrium structure.

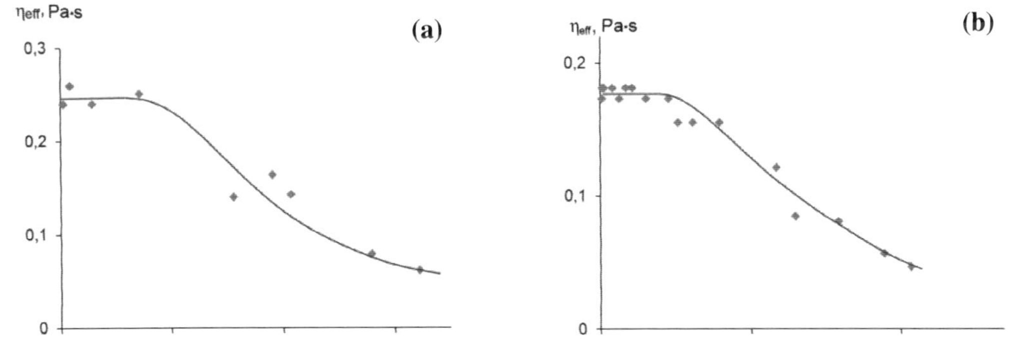

Figure 3. Effective viscosity versus shear stress: **(a)** SiO_2/PES-2, 50 nm; **(b)** SiO_2/PES-2, 100 nm.

5. Viscosity measurement at low flow rate gradients

An experimental setup has been created that makes it possible to study the flow of liquid in a capillary at low pressure differences at the ends of the capillary, when it can be expected that the liquid structure changes little during elementary acts of viscous flow. The use of conventional viscometers is impossible, because they fail to create extremely low flow velocity gradients. Figure 4a shows a diagram of the measuring setup. Liquid from one vessel flows to another through a capillary. The liquid level in Vessel 1 is determined by the shift in the reflection of the laser beam from its surface using the photodiode. The change in the level determines the rate of liquid flow through the capillary. The process of measuring is carried out in an automatic mode at a constant temperature.

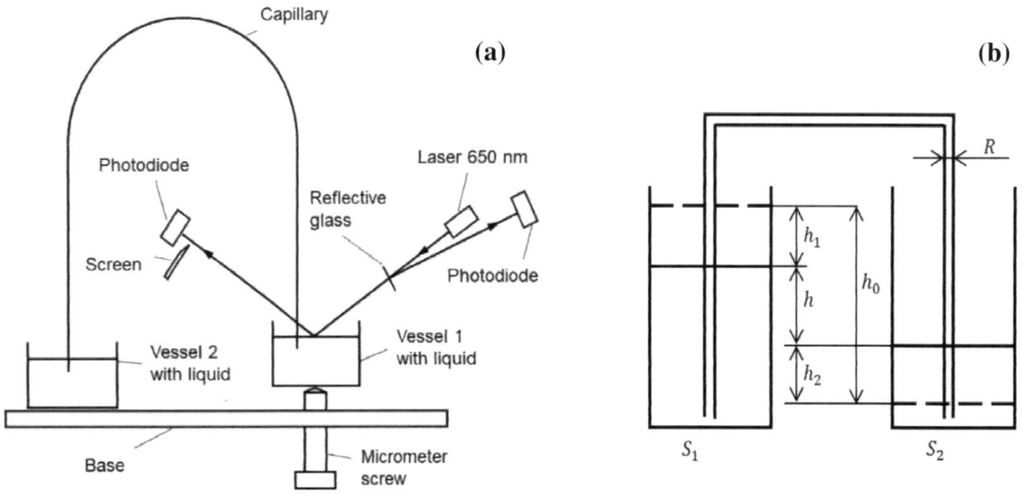

Figure 4. Measuring setup diagram **(a)** and scheme for measuring the level difference in vessels **(b)**.

A capillary connects two vessels with liquid with the initial difference in liquid levels h_0, the current liquid level difference is equal to $h = h_0 - h_1 - h_2$ (figure 4b). The pressure difference at the ends of the capillary is equal to the expression $\Delta p = \rho g h = \rho g(h_0 - h_1 S_1 \alpha)$, where ρ is the density of the liquid, S_1 and S_2 are the areas of the vessels and $\alpha = (1/S_1 + 1/S_2)$. When equilibrium is reached, the level difference will be equal to zero: $h_0 - h_{1,end} S_1 \alpha = 0$, where $h_{1,end}$ is the equilibrium liquid level of the in Vessel 1.

The volume v of the liquid flowing in $t = 1$ s through a cylindrical capillary of radius R and length l at a pressure difference at its ends Δp is [12]:

$$v = \frac{\pi R^4}{8\eta l}\Delta p = \frac{\pi \rho g R^4}{8\eta l} h, \tag{2}$$

where η is the dynamic viscosity of the liquid. Denoting $\gamma = \pi \rho g R^4 \alpha/(8\eta l)$, after transformations we obtain

$$\ln\left(h_{1,end}\left(h_{1,end} - h_1\right)^{-1}\right) = \gamma t. \tag{3}$$

From expression (3) it follows that if the viscosity of the liquid at any pressure difference, or at all gradients of the flow velocity remains constant, then the dependence of $\ln\left(h_{1,end}\left(h_{1,end} - h_1\right)^{-1}\right)$ on time t should be linear. If the viscosity increases at low gradients of the flow rate, then a deviation from this dependence will be observed. The absence of the effect of evaporation was checked by liquid level conservation in the Vessel 1 in the absence of a capillary for a period equal to the measurement time.

Figure 5a shows the dependences $\ln\left(h_{1,end}\left(h_{1,end} - h_1\right)^{-1}\right)$ on time t for PES-2 and a suspension of nanoparticles of silicon dioxide T-05 with a concentration of 1.25 wt.%. In the experiment, a capillary with a radius of 0.5 mm and a length of 955 mm was used for PES-2, and a capillary with a radius of 0.775 mm and a length of 174 mm was used for the suspension. It can be seen from the figure that at high gradients of the flow velocity, the dependence is linear. However, starting from a certain value of the level difference, for the suspension and for the base liquid increasing deviations from the linear dependence are observed, which indicates an increase in the viscosity.

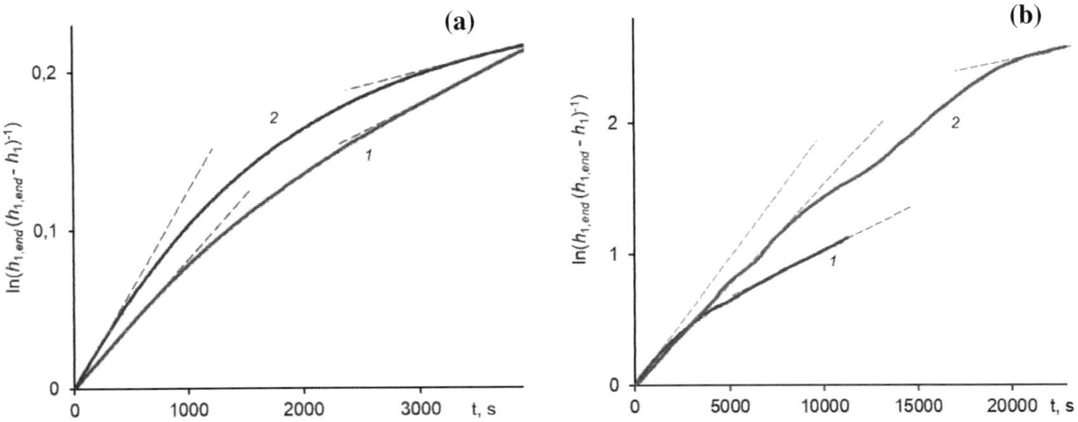

Figure 5. $\ln\left(h_{1,end}\left(h_{1,end} - h_1\right)^{-1}\right)$ vs. time *t* for: **(a)** *1* - PES-2; *2* - suspension SiO_2/PES-2, c = 1.25 wt.%; **(b)** *1* - castor oil; *2* - suspension SiO_2/castor oil, c = 1.25 wt.%.

It can be seen from formula (3) that the viscosity is inversely proportional to the tangent of the slope of the curve, figure 5a. Thus, the viscosity $\eta = \eta_0 \tan\varphi_0 (\tan\varphi)^{-1}$, where η_0 and $\tan\varphi_0$ are the initial viscosity and the tangent of curve slope angle respectively. As the gradient of the flow velocity decreases by the end of the experiment, the viscosity increases by two times for PES-2 and by ~ 6.2 times for the suspension.

To measure the increased viscosity of castor oil, which was chosen as a polar liquid, a capillary of 765 mm length and 3.01 mm in diameter was used. The viscosity of castor oil measured with the capillary viscometer VPZh-2 is 0.98 Pa·s. Figure 5b shows the change in the viscosity of castor oil and suspension of T-05 silicon dioxide particles in castor oil with a concentration of 1.25 wt.%. By the end of the experiment lasting more than six hours, the viscosity of castor oil increased by 2.7 times, for suspension it increased by 4 times.

6. IR spectroscopy investigation

Spectrometric studies of functional groups of investigated liquids in the mid-infrared region were carried out. The Bruker FTIR Spectrometer ALPHA obtained the IR spectra of suspensions of SiO_2/PES-2 and SiO_2/castor oil, nanoparticles of silica dioxide and hydrophobic base liquids.

Characteristic absorption peaks of siloxane and silanol groups are found in the IR spectra of silica dioxide nanoparticles, which are explained by the presence of OH groups on the particle surface. In the spectra of SiO_2/PES-2 suspensions, some changes are observed in comparison with the control spectrum of the base liquid, strengthening of the valence bonds Si–C and C–H, which indicate the interaction of nanoparticles with molecules of polyethylsiloxane liquid.

The figure 6a and the figure 6b show fragments of the IR spectrum of a SiO_2/PES-2 with particle concentration of 1.25 and 5 wt.% taking into account the background spectrum. Wide absorption bands were obtained in the frequency range associated with the absorption of the hydrogen bonds. At a higher concentration of nanoparticles in the suspension, the band is more pronounced.

The IR spectrum of castor oil shows a broad band in the region of hydrogen bond. In the IR spectrum of a suspension of SiO_2 particles in castor oil, there is also an absorption band in the region of OH group vibrations, associated with the formation of hydrogen bonds because of the interaction of nanoparticles and the base liquid, figure 6c.

7. Conclusion

Thus, the study of dynamic viscosity of colloidal suspensions of silica dioxide nanoparticles SiO_2 in polyethylsiloxane liquid (PES-2) and castor oil investigated by rotary and capillary methods showed the Newtonian behavior of the viscosity of studied suspensions. An increased value of the viscosity coefficients are found at small shear deformation angles and low flow rate gradients, which indicate the structurization of the studied liquids with the formation of a supramolecular equilibrium structure. With a decrease in the flow rate gradient, the viscosity of the suspensions increases more strongly than the viscosity of the base liquids, which characterizes a higher degree of suspension structuring by the end of the experiment. Moreover, the effect is more pronounced for a polar base liquid.

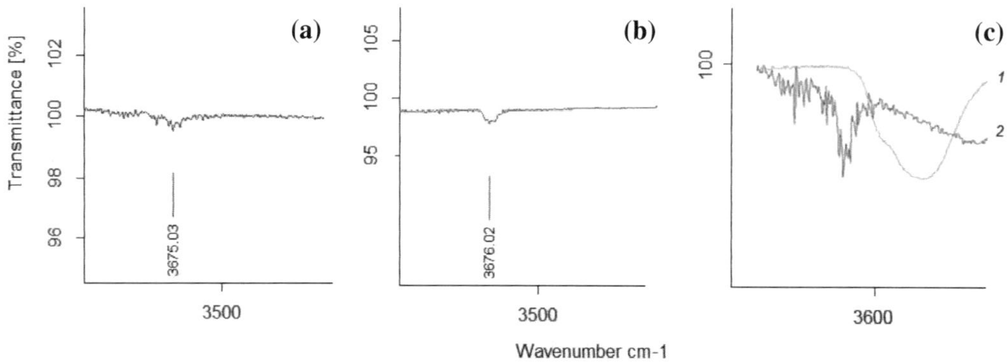

Figure 6. IR spectra of: **(a)** SiO_2/PES-2, c = 1.25 wt.%; **(b)** SiO_2/PES-2, c = 5 wt.%; **(c)** *1* - castor oil; *2* - suspension SiO_2/castor oil, c = 1.25 wt.%.

Spectrometric studies of suspensions of silica dioxide nanoparticles in polyethylsiloxane liquid and castor oil revealed the presence of absorption bands in the region of hydrogen bond vibrations that form a supramolecular structure. With an increase in the liquid flow rate gradient, the bonds that determine the supramolecular structure are destroyed like thixotropic structures under mechanical action. It leads to viscosity values measured by traditional methods, which can be assumed to refer to a liquid with a destroyed spatial structure.

The considered phenomenon of increased viscosity can be important in all processes where slow flows prevail, for example, when filtering liquids through artificial and natural membranes.

Acknowledgment
This work was partially supported by the RFBR grant No. 18-48-030020 r_a.

References

[1] Patent 2067077 Russian Federation, IPC C01B33/18
[2] Bardakhanov S P, Vikulina L S, Lysenko V I, Nomoev A V, Poluyanov S A and Tuzikov F V 2012 *Vestnik NSU. Series: Physics.* **7(4)** 107-16
[3] Rudyak V Y, Dimov S V and Kuznetsov V V 2013 *Tech. Phys. Lett.* **39** 779-82
[4] Namburu P K, Kulkarni D P, Dandekar A and Das D K 2007 *Micro Nano Lett.* **2** 67–71
[5] Badmaev B B, Bazaron U B, Derjaguin B V and Budaev O R 1983 Physica B+C 122 241–5
[6] Badmaev B, Dembelova T, Damdinov B, Makarova D and Budaev O 2011 *Colloids Surf.* A **383** 90–4
[7] Badmaev B B, Dembelova T S and Damdinov B B 2013 *Viscoelastic properties of polymer liquids* (Ulan-Ude: Publisher of the Buryat Scientific Center, SB RAS)
[8] Dembelova T S *et al* 2019 *IOP Conf. Ser.: Mater. Sci. Eng.* **704** 012005
[9] Badmaev B B, Dembelova T S, Makarova D N and Gulgenov C Z 2017 *Tech. Phys.* **62** 14–7
[10] Badmaev B B, Makarova D N, Damdinov B B, Sanditov D S and Dembelova T S 2014 *Russ. Phys. J.* **57** 742-8
[11] Dembelova T, Damdinov B, Baloshin Y, Barnakov Y and Petranovskii V 2018 *Solid State Phenomena* **271** 119-23
[12] Landau L D, Lifshitz E M 1986 *Theoretical Physic. Fluid Mechanics* (Moscow: Nauka)

IOP Conf. Series: Materials Science and Engineering 1000 (2020) 012006 doi:10.1088/1757-899X/1000/1/012006

Determination of nanoscale layer thickness adsorbed water liquid by acoustoelectric methods

I G Simakov, Ch Zh Gulgenov and S B Bazarova*

Institute of Physical Materials Science, Siberian Branch of the Russian Academy of Sciences, 670047, Ulan-Ude, Russia

*E-mail: baz_say@mail.ru

Abstract. It is shown to determine the thickness of the adsorbed liquid layer, it is proposed to use surface acoustic waves (SAW). The relative change in the SAW velocity is proportional to the thickness of the adsorption layer. The constant of proportionality is depending on the acoustic parameters of the solid (adsorbent) and the liquid in the layer. The velocity of sound in an adsorbed liquid depends on the thickness of the layer. For the determination of the velocity of sound in the liquid of the adsorption layer, it is proposed to use an original method based on comparing the parameters of bi-directional (differing in velocity) SAW.

Liquid layers are an adsorption layer, a wetting film or a bulk liquid layer. Surface acoustic waves (SAW) can be used to determine the thickness of the liquid layer. The dispersion of SAW in a layered system "piezoelectric substrate - liquid layer" is described by a large system of equations and can be solved only by numerical methods. The dispersion coefficient of the velocity of SAW in the system "piezoelectric substrate - thin layer of liquid" is determined by the expression:

$$\Delta V / V_S = \mathrm{Re}\left(\Delta k / k\right),$$

where Δk and ΔV are changes in the wavenumber and velocity of SAW as a result of acoustoelectric interaction

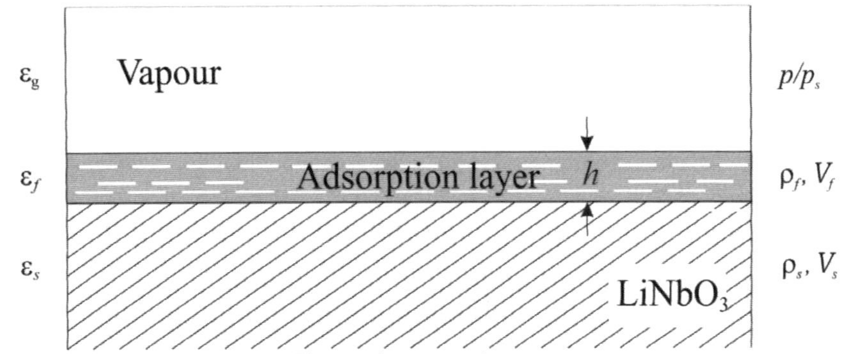

Figure 1. Layered system.

The dispersion of the velocity of SAW in the system "thin layer of liquid - piezoelectric substrate" (Figure 1) is described by dispersion equation (1). In [1] it has been shown that by linear approximation the relative change in velocity SAW ($\Delta V/V_s$) can be represented by a simple expression:

$$\frac{V_0 - V_s}{V_s} = \left(\frac{\rho_f V_f}{\rho_s V_s} \cdot \mathrm{tg}\,\theta + \pi K^2 \frac{\varepsilon_p\, \varepsilon'}{(\varepsilon_p + 1)^2} \right) \frac{h}{\lambda}, \tag{1}$$

where ρ_f and ρ_s – liquid density and solid density, ε_p, ε_f – relative permittivity's of $\varepsilon_p = (\varepsilon_{yy}\, \varepsilon_{zz} - \varepsilon_{yz}^2)^{1/2}$ and liquid, $\theta = \arccos (V_f/V_s)$ – the angle at which sound waves propagate in the liquid layer, V_f – sound velocity in liquid, V_s – surface wave velocity, K – electromechanical coupling coefficient. The variation in velocity of SAW depends on the acoustic and dielectric characteristics of the layered system.

For bulk liquids, linear approximation of the velocity variance of SAW, provided ($h << \lambda_s$), allows the determination of the thickness of the liquid layer from equation (1).

$$h = \lambda \cdot \frac{(\Delta V/V_s)}{\left(\frac{\rho_f V_f}{\rho_s V_s} \cdot \mathrm{tg}\,\theta + \pi K^2 \frac{\varepsilon_p\, \varepsilon'}{(\varepsilon_p + 1)^2} \right)},$$

In a vapor-gas medium, a nanoscale layer of adsorbed liquid on the surface of a solid is formed. The structure and physical properties of adsorbed liquid and liquid in the bulk state are different. Viscosity and permittivity are the most sensitive properties to changes in structure of liquid.

To exclude electrical disturbances in the propagation conditions of SAW, a piezoelectric with a metallized surface can be used. In this case, the expression for changing the velocity (1) is simplified

$$\frac{V_m - V_s}{V_s} = \frac{\rho_f V_f}{\rho_s V_s} \frac{h}{\lambda} \mathrm{tg}\,\theta = \frac{\rho_f}{\rho_s} \sqrt{1 - \frac{V_f^2}{V_s^2}} \frac{h}{\lambda}. \tag{2}$$

In this equation, the thickness of the adsorption layer determines the density of the liquid ρ_f and the sound velocity in the liquid V_f, which is determined from acoustic measurements.

To determine the velocity of sound in a liquid, the propagation of SAW in two different directions of an anisotropic metallized acoustic sound duct was used. The velocity of SAW depends on the direction of propagation. Therefore, it is possible to measure the relative change in velocity, which is caused by the influence of the liquid layer for different directions of the metallized sound line. The resulting system of linear equations determines V_f. For two directions of propagation of surface acoustic waves in a layered system with different values of the velocity ($V_1 \neq V_2$), we have two equations.

$$\frac{\Delta V_1}{V_1} = \frac{\rho_f}{\rho_s} \sqrt{1 - \frac{V_f^2}{V_1^2}} \frac{h}{\lambda}, \quad \frac{\Delta V_2}{V_2} = \frac{\rho_f}{\rho_s} \sqrt{1 - \frac{V_f^2}{V_2^2}} \frac{h}{\lambda}. \tag{3}$$

For different directions of SAW propagation in a layered system, the ratios ρ_f/ρ_s and h/λ are the same, and only the velocities V_1 and V_2 differ. From equations (3) we can write

$$\frac{V_2 \Delta V_1}{\Delta V_2 V_1} = \frac{\delta f_1}{\delta f_2} = \frac{V_2}{V_1} \frac{\sqrt{V_1^2 - V_f^2}}{\sqrt{V_2^2 - V_f^2}} \tag{4}$$

$\Delta V_i/V_i = \Delta f_i/f_i = \delta f_i$ from the interference condition ($i = 1, 2$).

$$V_f = \sqrt{\frac{R V_2^2 - V_1^2}{R - 1}}, \quad R = \left(\frac{\delta f_1 \cdot V_1}{\delta f_2 \cdot V_2} \right)^2 \tag{5}$$

Equation (5) makes it possible to calculate the adiabatic velocity of sound in an adsorbed liquid based on a comparison of the parameters of SAW differing in velocity in a layered system.

The position of the SAW – transducers used in this technique is shown in Figure 2. Four transducers *2* and *3*, deposited on the surface of the piezoelectric substrate 1, form two delay lines. Polymolecular adsorption of polar liquid vapor out on metal film 4 is carried. Which is applied between the emitting and receiving interdigital transducers. The dotted line marks the zone of interaction of SAW with the adsorption layer.

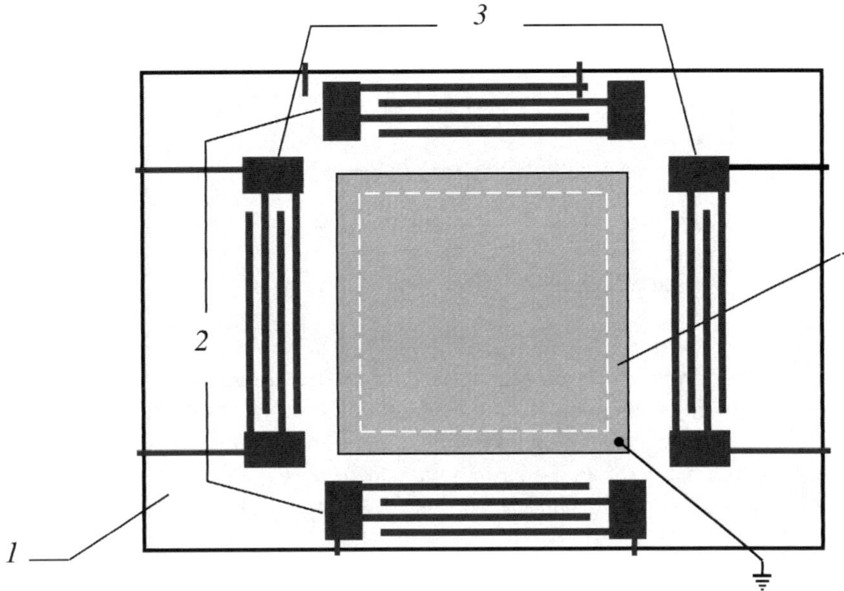

Figure 2. The position of the SAW – transducers

To determine the change in the velocity of surface acoustic waves in a layered system, a phase-interference method is used to measure small changes in the velocity and attenuation of surface acoustic waves [2].

The density of liquid in the layer can be calculated by the formula

$$\rho_f = \frac{K}{2V_s^2} + \sqrt{\left(\frac{K}{2V_s^2}\right)^2 + \left(\frac{\Delta V \rho_s \lambda}{V_s h}\right)^2}. \quad (6)$$

the resulting expression (2) taking into account the dependence of the sound velocity in the liquid $V_f^2 = 1/(\rho_f \beta)$ on its density and adiabatic compressibility β ($E = 1/\beta$ – bulk modulus of water).

To determine the density of a liquid using formula (6), it is sufficient to measure the relative change in the velocity of SAW in any of the selected directions of propagation of SAW ($V_s = V$ or $V_s = V_2$). The sensitivity of the interference method to a change in velocity is high (~ 10^{-8} at a frequency of 100 MHz). Wavelength λ is given by the dimensions of the transducer. The bulk modulus of water of the liquid in the boundary layer can be taken the same as that of the bulk liquid.

The method for determining the thickness of a nanoscale liquid layer was tested when evaluating the adsorption isotherm h(p/ps)T of water vapor on the metallized surface of lithium niobate YZ-cut (Figure 3).

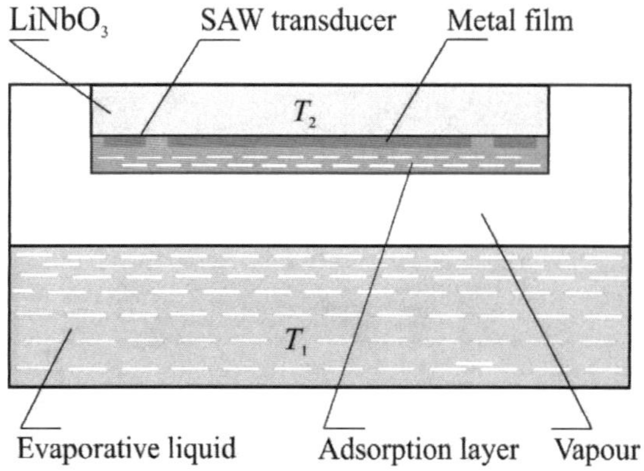

Figure 3. Diagram of the measuring cell.

Due to the use of tabular values of the bulk modulus of water, when determining the density of adsorbed water, it has an error due to structural differences. The compressibility of water in the boundary phase differs little from the compressibility of water in the bulk phase. Therefore, to determine the density of adsorbed water, one can use the value of the bulk modulus of water.

Numerical methods showed that with an increase in the density of bulk water under the influence of pressure from 0 to 2%, an increase in the bulk modulus in water is 0 – 12.8 %, an increase in the velocity of sound 0 – 6.05 %. The density, which is determined by formula (6), in addition to the dependence on the layer thickness and the bulk modulus, also depends on the parameters of SAW. With an increase in the density of adsorbed water to 2%, the error in determining the density increases only to 0.28%. When using the volumetric value of E, for example, in the well-known equality $\left(E = \rho_f V_f^2\right)$, an error appears. When using the bulk value of E in equation (6), the resulting error in determining the density δ is less than in the first case by about 10 times. The values of the variation in the acoustic parameters of bulk water during its compaction, as well as the error in determining the density δ are given in the table. $\Delta x = x(p) - x_0$, where x equal to ρ_f, E, V_f. Compaction occurs under the influence of pressure $\Delta p = (p - 1)$ atm.

The bulk modulus can be calculated using an empirical formula, which was obtained on the basis of numerical calculations of changes in density and other acoustic parameters of water under the influence of external pressure (table) [3]:

Table 1. Acoustic parameters of water during compaction

$\Delta\rho/\rho_f$, %	$\Delta E/E$, %	$\Delta V/V_f$, %	Δp, atm	δ, %
0,5	3,14	1,35	112	0,09
1,0	6,43	2,87	232	0,16
1,5	9,55	4,36	360	0,23
2,0	12,8	6,05	497	0,28

$$\Delta E / E \cong 6{,}4\, \Delta\rho / \rho_f \quad (7)$$

To minimize the error, it is necessary according to equation (6), determine $\Delta\rho/\rho_f$ using the tabular value of E for bulk water. The found $\Delta\rho/\rho_f$ value can be used in relation (7) to detail the bulk modulus and re-determine the density change.

To determine the thickness of the layer of adsorbed liquid, you can use one of the expressions obtained from equations (3).

$$h=\frac{\lambda\cdot\rho_s\cdot\delta f_1\cdot V_1}{\rho_f\sqrt{V_1^2-V_f^2}},\qquad h=\frac{\lambda\cdot\rho_s\cdot\delta f_2\cdot V_2}{\rho_f\sqrt{V_2^2-V_f^2}}\tag{8}$$

Accordingly, it is necessary to determine the density of the liquid ρ_f and the speed of sound in it V_f from acoustic measurements using formulas (3) – (6).

Thus, it is possible to determine the thickness of the nanoscale layer of adsorbed liquid from acoustic measurements of changes in the parameters of sounding pulses of SAW. An expression for determining the thickness of the adsorption layer is obtained. The proportionality factor depends on the acoustic parameters of the solid (adsorbent) and liquid in the layer. The velocity of sound in adsorbed liquid depends on the thickness of the layer. In order to detail the velocity of sound in the liquid of the adsorption layer, it is proposed to use an original method based on comparing the parameters of surface acoustic waves which differ in velocity.

Acknowledgment

The work was performed under the IPMS SB RAS state budget project №0336-2019-0010.

References

[1] Simakov I G 2014 *Uchenyye zapiski fizicheskogo fakulteta Moskovskogo universiteta* **5** 145325.
[2] Simakov I G, Gulgenov Ch Zh 2011 *Vestnik Buryatskogo gosudarstvennogo universiteta. Khimiya. Fizika* **3** 216–220.
[3] Simakov I G, Gulgenov Ch Zh 2019 Akustoelektricheskie metody issledovaniya zhidkostej v granichnoj faze 222 s.

T-x-y-z Diagram Prediction for the Quaternary System Li,Na,Ca,La||F

V Vorob'eva, A Zelenaya, V Lutsyk and **M Lamueva**

Institute of Physical Materials Science (Siberian Branch of Russian Academy of Sciences), Ulan-Ude, Russian Federation

E-mail: vvorobjeva@mail.ru

Abstract. The prediction of the geometric structure of the LiF-NaF-CaF_2-LaF_3 quaternary system T-x-y-z diagram has been carried out on the basis the data about the binary systems and the 3D computer models of the T-x-y diagrams of boundary ternary systems, designed earlier; and its prototype has been constructed. Correctness of the prediction may be confirmed (or rejected) by the experiments only.

1. Introduction

In the literature are given the diverse Variants of the potential fuel of IV generation nuclear reactors, both calculated thermodynamically and studied experimentally, are given in literature [1-7]. However, the represented information about the geometrical structure of the corresponding phase diagrams is limited only by the visualization of the liquidus surfaces and (not always) a certain number of isothermal sections and isopleths. Whereas it is necessary for obtaining the information about the results of interaction of initial components the complete model of diagram, which includes phase regions with the liquid and solid-phase regions.

The developed 3D and 4D computer models of T-x-y and T-x-y-z diagrams of ternary and quaternary systems, correspondingly, serve both for the visualization of their geometrical structure and are the source of obtaining information about the crystallization stages and formation of microstructure, give possible to qualitatively and quantitatively evaluate the relations of phases for the assigned composition [8, 9]. This complex study of the fluoride systems properties gives the possibility to optimize the selection of fuel and to predict its properties, and also is the theoretical basis for the experimental investigations.

The compositions, obtained from the LiF-NaF-CaF_2-LaF_3 system тоже are investigated in connection with the development of IV generation nuclear reactors [1]. Fluoride of lanthanum (LaF_3) serves as a proxy-compound for fluoride of plutonium (PuF_3), since the direct use of fluoride of plutonium (PuF_3) would cause enormous experimental difficulties.

2. 3D computer models of T-x-y diagrams for systems forming four-component system of lithium, sodium, calcium and lanthanum fluorides

The surfaces of the phase diagrams formed by the fluorides of lithium, sodium, calcium, and lanthanum are not complicated. But the geometric structure of these diagrams becomes more complicated by the formation of the binary $NaLaF_4$ compound and polymorphism of CaF_2 [4].

Before designing a computer 3D (and 4D too) model, it is necessary to discuss the boundary systems, as well as to make a formal indication of the considered quaternary system, that is, to indicate the LiF-NaF-CaF_2-LaF_3 system as A-B-C-D (Fig. 1), and to give the necessary indication of all phase transformations [10]. For example, two polymorphic modifications of CaF_2 are involved in the formation of the LiF-NaF-CaF_2-LaF_3 (A-B-C-D) diagram and they received the appropriate indicators: C and C1.

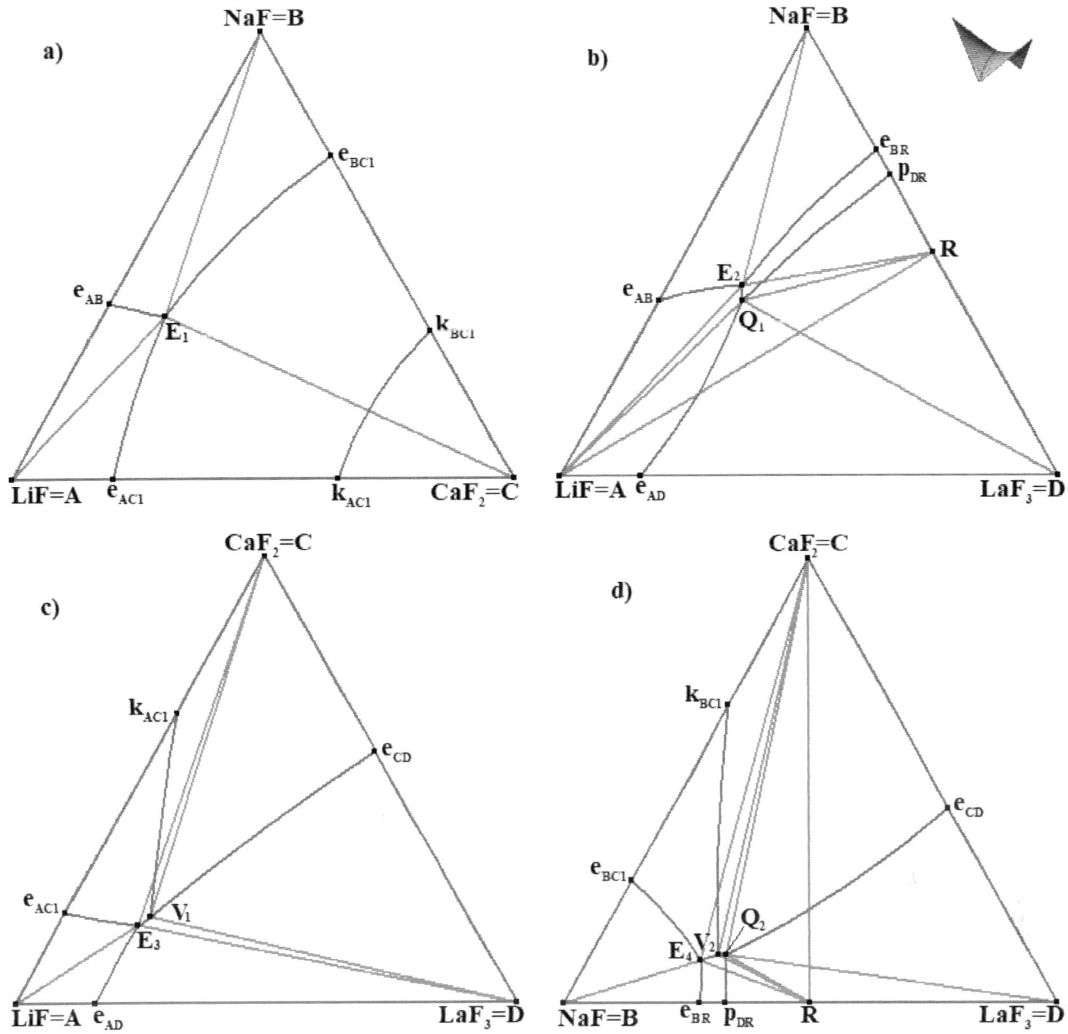

Figure 1. 3D компьютерные модели T-x-y диаграмм LiF-NaF-CaF_2 (A-B-C) (a), LiF-NaF-LaF_3 (A-B-D) (b), LiF-CaF_2-LaF_3 (A-C-D) (c), NaF-CaF_2-LaF_3 (B-C-D) (d).

2.1. T-x diagrams of binary systems

The systems LiF-NaF (A-B), LiF-LaF_3 (A-D) and CaF_2-LaF_3 (C-D) are the eutectic ones, but the last is characterized more over by the eutectoid reaction C→C1+D because of the polymorphism of calcium fluoride.

The LiF-CaF_2 (A-C) and NaF-CaF_2 (B-C) systems are also of eutectic type, but the allotropy of calcium fluoride (CaF_2) is the reason of the polymorphic transition in the form of the metatectic reaction (C→C1+L).

The NaF-LaF_3 (B-D) system has a typical T-x diagram with the peritectic p_{DR}: L+D→R reaction of forming the incongruently melting $NaLaF_4$ (R) compound and eutectic e_{BR}: L→B+R reaction.

2.2. T-x-y diagrams of ternary systems

The LiF-NaF-CaF_2 (A-B-C) ternary system is characterized by eutectic invariant reaction E_1: L→A+B+C1 and the univariant polymorphic C→C1+L transition between two CaF_2 modifications.

The LiF-NaF-LaF_3 (A-B-D) system with the $NaLaF_4$ (R) binary incongruently melting compound includes the quasi-peritectic Q_1: L+D→A+R and eutectic E_2: L→A+B+R invariant reactions.

It seems that the LiF-CaF_2-LaF_3 (A-C-D) and the LiF-NaF-LaF_3 (A-B-D) systems are similar. Except the eutectic reaction E_3: L→A+D+C1 the A-B-D system includes one additional invariant reaction. It is called a quasi-peritectic reaction in [1] and is written as L+C→C1+D. However, it is not. Because of the polymorphous transfer in the LiF-CaF_2 (A-C) binary system the V_1: C→C1+D+L polymorphic transition also takes a place with the passive role of the L and LaF_3 (D). More details on ternary systems with allotropy of components, including a similar case, are described in the paper [11].

Construction of the 3D computer model of the T-x-y diagram begins with the 3D scheme of uni- and invariant states [11, 12]. This is a 3D variant of well-known scheme of phase reactions (planar Sheil scheme) [13], supplemented by the concentrations of phases, which participate in three-phase transformations. This very important addition opens wide possibilities for describing all surfaces and phase regions of the T-x-y diagram (as it is shown in previous slides for the system, formed by fluorides of sodium, calcium and lanthanum.

The polymorphic transition V_2: C→C1+R+L is intermediate between two invariant reactions: the quasi-peritectic Q_2: L+D→C+R and the eutectic E_4: L→B+R+C1 in the NaF-CaF_2-LaF_3 (B-C-D) system. So, its scheme leads to the third invariant reaction E_5: C→C1+D+R of the eutectoid type.

The scheme of uni- and invariant states allows to define all unruled ruled surfaces, horizontal planes, according to invariant reactions, and all phase regions with liquid and without liquid.

Thus, a 3D computer model provides an excellent opportunity to get a complete and comprehensive view of the T-x-y diagram, either its geometric structure or the options for its melts crystallization. In addition, 3D models are very useful as for their structures understanding and for the 4D model designing of the four-component system T-x-y-z diagram, formed by them.

Each spatial computer model is a prototype of a phase diagram, which is able to become a perfect model of a real system after the experimental or calculated parameters (concentrations and temperatures of binary and ternary points, curvature characteristics of surfaces, according to isothermal sections and isopleths) input.

3. 4D computer model of the predicted T-x-y-z diagram

The same approach is applied to design the 4D computer model of the T-x-y-z diagram. As for the LiF-NaF-CaF_2-LaF_3 (A-B-C-D) system, the logic of the bi-, uni- and invariant state schemes leads to the only possible scheme of the phase reactions of the quaternary system (Table 1) [14].

From the scheme it follows those three invariant reactions can be expected the polymorphic transition CaF_2→CaF_2'+LaF_3+$NaLaF_4$+L (or ν: C→C1+D+R+L) between two modifications of calcium fluoride (C и C1) in the presence of liquid L, LaF_3 (D) and $NaLaF_4$ (R) compound at temperatures below 985 K, the quasi-peritectic reaction L+LaF_3→LiF+CaF_2'+$NaLaF_4$ (π: L+D→A+C1+R) below 869 K and the eutectic reaction L→LiF+NaF+CaF_2'+$NaLaF_4$ (ε: L→A+B+C1+R) below 854 K (Fig. 2-3).

In general, the T-x-y-z diagram includes:6 hypersurfaces of liquidus and 6 - of solidus ones; 11 pairs of solvus hypersurfaces (total - 22); 2 hypersurfaces of the transus; 22 triads of ruled hypersurfaces with a generated line (66 in total); 52 ruled hypersurfaces with a generated plane; 3 complexes, corresponding to the invariant transformations ν, π, ε, each of them consists of five horizontal (isothermal) hyperplanes.

All these 169 hypersurfaces serve as the boundaries of 62 phase regions: 6 single-phase regions I (A, B, C, C1, D, R) and 6 two-phase regions L+I; 12 two-phase regions I+J without liquid (A+B, A+C1, A+D, A+R, B+C1, B+R, C+D, C+R, C+C1, C1+D, C1+R, D+R) and 12 three-phase regions with liquid L+I+J; 10 three-phase regions I+J+K without liquid (A+B+C1, A+ B+R, A+C1+D, A+C1+R, A+D+R, B+C1+R, C+C1+D, C+C1+R, C+D+R, C1+D+R) and 10 four-phase regions with liquid L+I+J+K; 3 four-phase regions without liquid A+B+C1+R, A+C1+D+R, C+C1+D+R, and 3 five-phase regions, degenerated into planar hyperplanes, L+A+B+C1+R, L+A+C1+D+R, L+C+C1+D+R.

IOP Conf. Series: Materials Science and Engineering 1000 (2020) 012007 doi:10.1088/1757-899X/1000/1/012007

Table 4. The phase reactions scheme of the LiF-NaF-CaF_2-LaF_3 (A-B-C-D) T-x-y-z diagram with the incongruently melting compound $NaLaF_4$ (R) (Fig. 2-3), $D>C>e_{CD}>k_{AC1}=k_{BC1}>B>A>e_{BC1}>p_{DR}>e_{AD}>e_{AC1}>Q_2>V_2>e_{BR}>V_1>e_{C1D}>E_4>E_3>e_{AB}>E_1>Q_1>E_2>E_5>\nu>\pi>\varepsilon$

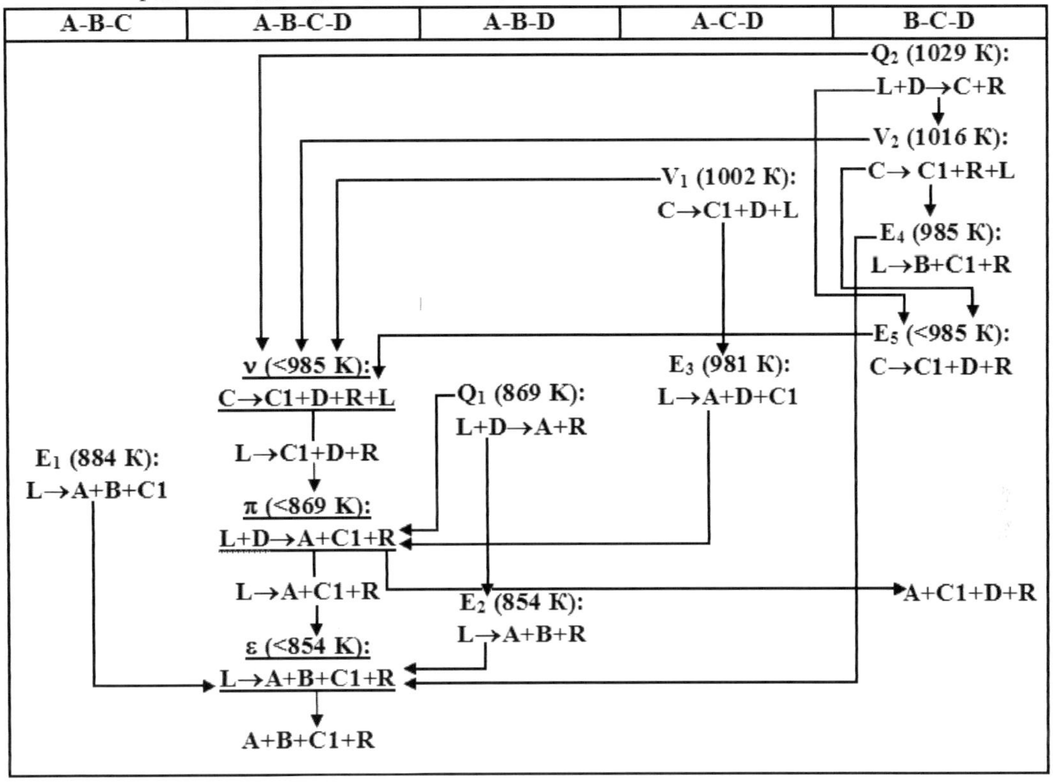

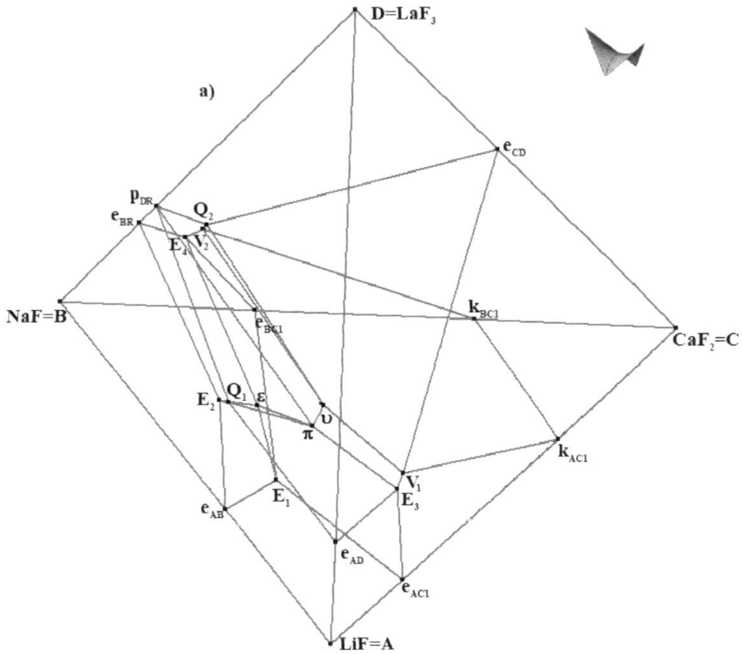

Figure 2. Liquidus prototype 4D model: x-y-z projection of the LiF-NaF-CaF_2-LaF_3 (A-B-C-D) T-x-y-z diagram.

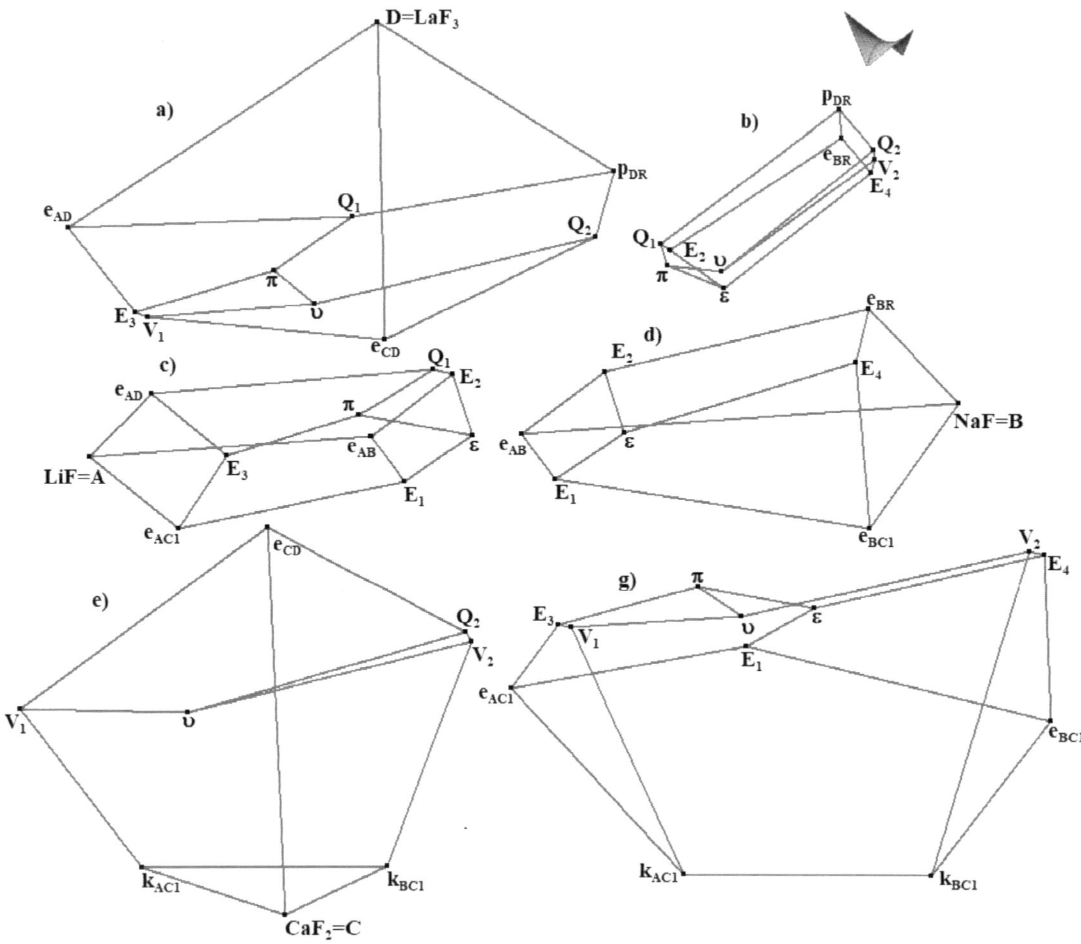

Figure 3. Liquidus prototype 4D model: x-y-z projection of the hypersurfaces D (b), R (c), A (d), B (e), C (f), C1 (g) separately.

4. Summary

The T-x-y-z diagram, based on the T-x-y diagrams 3D models [14], has been predicted and the 4D model of its prototype has been constructed. It includes the polymorphic transition between two modifications of calcium fluoride with the passive role of liquid, LaF_3 and the $NaLaF_4$ compound, the quasi-peritectic and eutectic invariant reactions. In general, the T-x-y-z diagram may consist of 169 hypersurfaces and 66 phase regions.

When designing spatial computer models of phase diagrams of ternary, quaternary, and more complex systems, it should be paid more attention to updating the data on their edges, including information about the initial components and compounds. In the 4D model of the LiF-NaF-CaF_2-LaF_3 diagram, considered in this paper and forming it 3D models of ternary systems, it is necessary in the future to take into account the decomposition of the binary compound $NaLaF_4$ in the sub-solidus at 330°C, which is not mentioned in [15], where a binary system has been limited from below by the temperature of 600 K.

This 4D model is based on data from [1] with CaF_2 allotropy. However, this property of calcium fluoride is not taken into account in [10-12]. It is possible that the authors of these papers [15-17]. It is possible that the authors of these papers were not interested, for practical reasons, in such high temperatures (1420 K) and they did not take into account the high-temperature modification of calcium fluoride, stable at atmospheric pressure [18].

Acknowledgement
This work has been performed under the program of fundamental research SB RAS (project 0336-2019-0008) and was partially supported by the Russian Foundation for Basic Research projects № 19-38-90035, 20-21-00056.

References

[1] Beneš O and Konings R J M 2010 *Comprehensive Nuclear Materials* (Elsevier Ltd) **3** 359; doi: 10.1016/B978-0-08-056033-5.00062-8.
[2] Meer J P M, Konings R J M and Oonk H A J 2006 *J Nuclear Materials* **357**(1) 48; doi: 10.1016/j.jnucmat.2006.05.042.
[3] Beneš O and Konings R J M 2008 *J Nuclear Materials* **375**(2) 202; doi: 10.1016/j.jnucmat.2008.01.007.
[4] Beneš O and Konings R J M 2008 *J Nuclear Materials* **377**(3) 449; doi: 10.1016/j.jnucmat.2008.04.004.
[5] Beneš O, Meer J P M and Konings R J M 2007 *CALPHAD* **31**(2) 209; doi:10.1016/j.calphad.2006.12.004.
[6] Beneš O and Konings R J M 2008 *CALPHAD* **32**(1) 121; doi:10.1016/j.calphad.2007.07.006.
[7] Capelli E, Beneš O and Konings R J M 2014 *J Nuclear Materials* **449**(1) 111; http://dx.doi.org/10.1016/j.jnucmat.2014.03.009.
[8] Lutsyk V I, Vorob'eva V P and Zelenaya A E 2019 *Russ J Phys Chem* **93**(13) 2593; doi: https://doi.org/10.1134/S0036024419130156.
[9] Lutsyk V I and Vorob'eva V P 2017 *Russ J Phys Chem* **91**(13) 2593; doi: 10.1134/S0036024417130131.
[10] Beilmann M, Beneš O, Konings R J M, Fanghänel Th 2011 *J Chem Thermodynamics* **43** 1515; doi: 10.1016/j.jct.2011.05.002.
[11] Lutsyk VI and Vorob'eva V P 2010 *J Therm Anal. Calorim* **101**(1) 25; doi: 10.1007/s10973-010-0855-0.
[12] Lutsyk V I and Vorob'eva V P 2016 *Rus J Inorgan Chem* **61**(2) 188; doi: 10.1134/S0036023616020121.
[13] Miura S 2006 *J Phase Equilibria and Diffusion* **27**(1) 34; doi: 10.1361/105497106X92781.
[14] Vorob'eva V P, Zelenaya A E, Lutsyk V I and Lamueva M V 2020 *Nanosystems: Physics, Chemistry, Mathematics* **11**(3) 345; doi: 10.17586/2220-8054-2020-11-3-345-354.
[15] Sobolev B P and Fedorov P P 1978 *J Less-Common Metals* **60** 33.
[16] Švantner M, Mariani E, Fedorov P P, et al. 1979 *Crystal Research and Technology* **14**(3) 365.
[17] Fedorov P P, Mayakova M N, Kuznetsov S V, et al 2016 *Rus J Inorgan Chem* **61**(11) 1472; doi: 10.1134/S003602361611005X.
[18] Cazorla C and Errandonea 2014 *Physical review letters* **113** 235902; doi: https://doi.org/10.1103/PhysRevLett.113.235902.

IOP Conf. Series: Materials Science and Engineering 1000 (2020) 012008 doi:10.1088/1757-899X/1000/1/012008

Material Balances Design in the Al_2Yb-Al-Al_4Sr Subsystem

V Lutsyk, A Zelenaya and **M Parfenova**

Institute of Physical Materials Science (Siberian Branch of Russian Academy of Sciences), Ulan-Ude, Russian Federation

E-mail: vluts@ipms.bscnet.ru

Abstract. The vertical mass balances - for the given center of masses, and the horizontal ones - for the isothermal state of isopleth, have been analyzed. These results are used for the different purposes: microstructures characterization, solidification paths confirmation, phase regions 3D-prototyping for the exploded diagram. It is once more showed, that a computer assembled model of the T-x-y diagram is a useful tool to solve different fundamental and applied tasks of the materials science.

1. Introduction

Like the purpose of the work "Phase Diagrams: The Beginning of Wisdom" [1] is to present a primer on ''How to Read and Apply Phase Diagrams'' in the current environment of powerful software packages, this paper illustrates the new opportunities, that have been arrived with the spacious computer models of the T-x-y and more complex diagrams for the multicomponent systems invention.

"Phase diagrams are the perfect road map to understand the conditions for phase formation or transformation in any material system caused by variation of temperature, composition, pressure or any other viable state variable. That is why one can use phase diagrams as the starting point for materials design and process optimization by manipulating composition and processing variables to achieve the desired microstructures. That applies to all sorts of materials, such as alloys, ceramics, semiconductors, cement, concrete etc., and to a multitude of processes, such as melting, casting, crystal growth, joining, solid-state reaction, heat treatment/phase transformation, oxidation, vapor deposition, and so on" [1]. Computer model of T-x-y diagram adds many new applications, thanks to the two kinds of material balances: the vertical ones – for the given center of masses, and the horizontal matbalances – for the isothermal states on the isopleth.

On the road to assemble a T-x-y diagram there are some obstacles. One of them is the initial description of the phase diagram. Let's discuss it in the next section of this paper.

2. Graphics correction within T-x-y diagrams with melt immiscibility.

In some cases the errors in the phase diagrams description prohibit a right understanding of their graphics. First example (Fig. 1, a-b) corresponds to the series of phase diagrams with the melt immiscibility in [2, p. 119-222]. One-type error is present in the description of some T-x-y diagrams: for the surface of solvus with 5 dots on its contour $a_Qa_Ca^0_Ca^0_Ea_E$ (Fig. 1, a) a line $a^0_Ca^0_E$ is lost (Fig. 1, b). As a result, a contour projection of this surface isn't closed. At the same time, the solvus surfaces are supposed to be of cylindrical type with the orthogonal position of their originating segment and project into a curve. Such description adds the difficulties in understanding of the diagrams structure. Surely a computer model of phase diagram permits to image all possible variants of degeneration for the surfaces.

In [3, p. 368], while imaging a T-x-y diagram (Fig. 1, c) with a complex structure of the immiscibility dome (nQkMm), a tie-line ka_k has been lost, that conjugates a critical dot of immiscibility k and the end dot a_k of the fold a_ka on the solidus surface $a_BACc_Baa_k$. This misleading was overcome only when the right picture has been found in the original monograph [3] in English (Rhines F. N. Phase Diagrams in Metallurgy. Their Development and Application. N.Y.-Toronto-London. 1956).

More often are the situations with the concentration coordinates correction for the subsystems of the more large system, like in this paper.

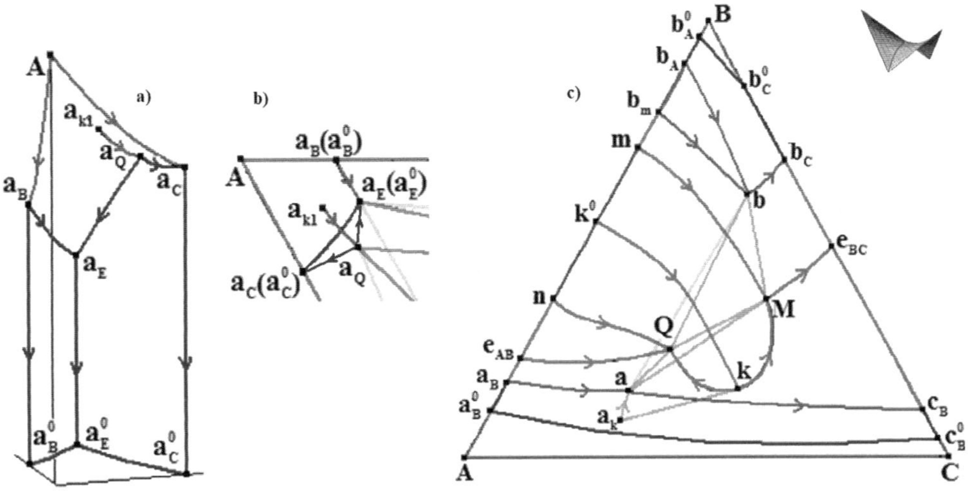

Figure 1. Fragment of diagram with a solvus surface v_{AB}, projecting into a line [2] (a, b); projection of diagram with a dome of liquid immiscibility crossing a eutectic curve $e_{AB}e_{BC}$ [3] (c)

3. 3D computer models of T-x-y diagrams for subsystem Al_2Yb-Al-Al_4Sr

The sub system Al_2Yb-Al-Al_4Sr was obtained by the dividing of ternary system Yb-Al-Sr with the quasi-binary section Al_2Yb-Al_4Sr.

As the initial data for simulation of model, we used information about the structure of binary systems concerning the coordinates of binary points and the resulting compounds. Systems Al-Sr and Al-Yb are presented in the reference book [4].

In this case, the subsystem Al-Al_4Sr contains one binary eutectic, the subsystem Al-Al_2Yb includes the eutectic, the peritectic and the incongruently melting compound Al_3Yb. The work [5] presents data on the quasi-binary section Al_2Yb-Al_4Sr (this subsystem is a simple eutectic), and also shows the liquidus of the subsystem Al_2Yb-Al-Al_4Sr, and two invariant points are defined: eutectic ($L_E \rightarrow Al+SrAl_4+YbAl_3$) and quasiperitectic ($L_Q+YbAl_2 \rightarrow SrAl_4+YbAl_3$) [5].

For the correct simulation of model, it is necessary to transform the coordinates of binary points from the systems Al-Sr and Al-Yb to the subsystems Al-Al_4Sr and Al-Al_2Yb using the matrix transformations: Z=M·x, where Z – vector column of point coordinates in the global system, M - matrix of coordinates of subsystem points in the global system, x - vector column of point coordinates in the subsystem [6-7].

So the eutectic point in the system Al-Sr near Al (e_{BC}) has coordinates (0.99; 0.01), and the compound Al_4Sr owns (0.8; 0.2). To convert its coordinates from the system Al-Sr to the subsystem Al-Al_4Sr, a matrix transformation is compiled:

$$e_{BC}:\begin{pmatrix}0{,}99\\0{,}01\end{pmatrix}=\begin{pmatrix}1 & 0{,}8\\0 & 0{,}2\end{pmatrix}\begin{pmatrix}x_1\\x_2\end{pmatrix}\rightarrow\begin{pmatrix}x_1\\x_2\end{pmatrix}=\begin{pmatrix}1 & 0{,}8\\0 & 0{,}2\end{pmatrix}^{-1}\begin{pmatrix}0{,}99\\0{,}01\end{pmatrix}\rightarrow\begin{pmatrix}x_1\\x_2\end{pmatrix}=\begin{pmatrix}0{,}95\\0{,}05\end{pmatrix}.$$

Thus, the coordinates of binary eutectic e_{BC} in the subsystem Al-Al_4Sr is (0.95; 0.05).

Similarly, the coordinates of eutectic (e_{BR}), peritectic (p_{AR}) and compound R=Al_3Yb are converted from the system Al-Yb to the subsystem Al-Al_2Yb:

$$e_{BR}:\begin{pmatrix}0{,}96\\0{,}04\end{pmatrix}=\begin{pmatrix}1 & 0{,}67\\0 & 0{,}33\end{pmatrix}\begin{pmatrix}x_1\\x_2\end{pmatrix}\rightarrow\begin{pmatrix}x_1\\x_2\end{pmatrix}=\begin{pmatrix}1 & 0{,}67\\0 & 0{,}33\end{pmatrix}^{-1}\begin{pmatrix}0{,}96\\0{,}04\end{pmatrix}\rightarrow\begin{pmatrix}x_1\\x_2\end{pmatrix}=\begin{pmatrix}0{,}879\\0{,}121\end{pmatrix}.$$

$$p_{AR}: \begin{pmatrix}0,87\\0,13\end{pmatrix} = \begin{pmatrix}1 & 0,67\\0 & 0,33\end{pmatrix}\begin{pmatrix}x_1\\x_2\end{pmatrix} \rightarrow \begin{pmatrix}x_1\\x_2\end{pmatrix} = \begin{pmatrix}1 & 0,67\\0 & 0,33\end{pmatrix}^{-1}\begin{pmatrix}0,87\\0,13\end{pmatrix} \rightarrow \begin{pmatrix}x_1\\x_2\end{pmatrix} = \begin{pmatrix}0,61\\0,39\end{pmatrix}.$$

$$R: \begin{pmatrix}0,75\\0,25\end{pmatrix} = \begin{pmatrix}1 & 0,67\\0 & 0,33\end{pmatrix}\begin{pmatrix}x_1\\x_2\end{pmatrix} \rightarrow \begin{pmatrix}x_1\\x_2\end{pmatrix} = \begin{pmatrix}1 & 0,67\\0 & 0,33\end{pmatrix}^{-1}\begin{pmatrix}0,75\\0,25\end{pmatrix} \rightarrow \begin{pmatrix}x_1\\x_2\end{pmatrix} = \begin{pmatrix}0,24\\0,76\end{pmatrix}.$$

It should be noted that in work [5], preliminary transformation of the binary points coordinates was not done at the studying of ternary subsystem Al_2Yb-Al-Al_4Sr. And the obtained ternary subsystem Al_2Yb-Al-Al_4Sr uses the coordinates of the Al-Sr and Al-Yb systems on the one side and the quasi-binary system Al_2Yb-Al_4Sr on the other. In this regard, for the correct model construction, the coordinates of ternary points were recalculated through the compositions of the local simplex Al_2Yb-Al-Al_4Sr:

$$Q: \begin{pmatrix}0.043\\0.84\\0.117\end{pmatrix} = \begin{pmatrix}0.33 & 0 & 0\\0.67 & 1 & 0.8\\0 & 0 & 0.2\end{pmatrix}\begin{pmatrix}x_1\\x_2\\x_3\end{pmatrix} \rightarrow \begin{pmatrix}x_1\\x_2\\x_3\end{pmatrix} = \begin{pmatrix}0,13\\0,285\\0,585\end{pmatrix},$$

$$E: \begin{pmatrix}0.032\\0.95\\0.018\end{pmatrix} = \begin{pmatrix}0.33 & 0 & 0\\0.67 & 1 & 0.8\\0 & 0 & 0.2\end{pmatrix}\begin{pmatrix}x_1\\x_2\\x_3\end{pmatrix} \rightarrow \begin{pmatrix}x_1\\x_2\\x_3\end{pmatrix} = \begin{pmatrix}0,1\\0,81\\0,09\end{pmatrix}.$$

Thus, the coordinates of points on the surfaces contour were transformed from the system Yb-Al-Sr to the subsystem Al-Al_2Yb-Al_4Sr (Table 1) and used for the construction of T-x-y diagram model. To simplify the model, the solid-phase solubility was not taken into account. The model was developed by the assembly method from phase regions [8-9] using the author's software [10].

The obtained computer model of T-x-y diagram for subsystem Al_2Yb-Al-Al_4Sr (A-B-C) contains four liquidus surfaces corresponding to the components Al_2Yb (A), Al (B), Al_4Sr (C) and compound R=Al_3Yb; ten ruled surfaces associated with them; four ruled surfaces at the boundary of the two-phase region R+C; two horizontal complexes at the temperatures of invariant points E and Q (Table 2); five two-phase regions (L+A, L+B, L+C, L+R, R+C) and seven three-phase regions (L+A+C, L+A+R, L+B+R, L+B+C , L+C+R, A+C+R, B+C+R) (Table 3, Fig. 2). The R+C phase region has a degenerate structure in the form of a vertical plane.

Table 1. Coordinates of initial points on the surfaces contour.

Точки	z1	z2	z3	T
A=Al_2Yb	1	0	0	1360
B=Al	0	1	0	660,425
C=Al_4Sr	0	0	1	1040
e_{AC}	0,13	0	0,87	1025
e_{BC}	0	0,95	0,05	654
p_{AR}	0,39	0,61	0	980
e_{BR}	0,121	0,879	0	625
E	0,1	0,81	0,09	620
Q	0,13	0,285	0,585	800
R=Al_3Yb	0,76	0,24	0	800
A_c	0,6	0	0,4	1025
C_A	0,03	0	0,907	1025

Table 2. Surfaces contours.

№	Name	Contour	№	Name	Contour
			Liquidus		
1	q_A	$Ap_{AR}Qe_{AC}$	3	q_C	$Ce_{AC}QEe_{BC}$
2	q_B	$Be_{BC}Ee_{BR}$	4	q_R	$p_{AR}QEe_{BR}$

Ruled surfaces					
5	q^r_{AR}	$p_{AR}QA_QA_R$	14	q^r_{RC}	EQR_QR_E
6	q^r_{RA}	$p_{AR}QR_QR_A$	15	s^r_{CR}	$C_ER_ER_QC_Q$
7	q^r_{AC}	$e_{AC}QA_QA_C$	16	v^R_{AC}	$A_QC_QC^0_QA^0_Q$
8	q^r_{CA}	$e_{AC}QC_QC_A$	17	v^r_{AR}	$A_QR_QR^0_QA^0_Q$
9	q^r_{BR}	$e_{BR}EB_EB_R$	18	v^r_{BC}	$B_EC_EC^0_EB^0_E$
10	q^r_{RB}	$e_{BR}ER_ER_B$	19	v^r_{BR}	$B_ER_ER^0_QB^0_E$
11	q^r_{BC}	$e_{BC}EB_EB_C$	20	$v^r_{CR(E)}$	$C_ER_ER^0_EC^0_E$
12	q^r_{CB}	$e_{BC}EC_EC_B$	21	$v^r_{CR(Q)}$	$C_QR_QR^0_QC^0_Q$
13	q^r_{CR}	EQC_QC_E	22		
Horizontal complexes					
23	h^E_{BCR}	$B_EC_ER_E$	27	h^Q_{ACR}	$A_QC_QR_Q$
24	h^E_{BCE}	B_EC_EE	28	h^q_{ACQ}	A_QC_QQ
25	h^E_{BRE}	B_ER_EE	29	h^Q_{ARQ}	A_QR_QQ
26	h^E_{CRE}	C_ER_EE	30	h^Q_{CRQ}	C_QR_QQ

Table 3. Structure of phase regions.

№	Phase region	Boundary surfaces	№	Phase region	Boundary surfaces
1	L+A	q_A, q^r_{AC}, q^r_{AR}	6	L+B+R	q^r_{BR}, q^r_{RB}, h^E_{BRE}
2	L+C	q_C, q^r_{CA}, q^r_{CR}, q^r_{CB}	7	L+B+C	q^r_{BC}, q^r_{CB}, h^E_{BCE}
3	L+B	q_B, q^r_{BR}, q^r_{BC}	8	L+C+R	q^r_{CR}, q^r_{RC}, h^Q_{CRQ}, h^E_{CRE}, s^r_{CR}
4	L+A+C	q^r_{AC}, q^r_{CA}, h^q_{ACQ}	9	A+C+R	h^Q_{ACR}, v^R_{AC}, v^r_{AR}, $v^R_{CR(Q)}$
5	L+A+R	q^r_{AR}, q^r_{RA}, h^Q_{ARQ}	10	B+C+R	h^E_{BCR}, v^r_{BR}, v^r_{BC}, $v^R_{CR(E)}$

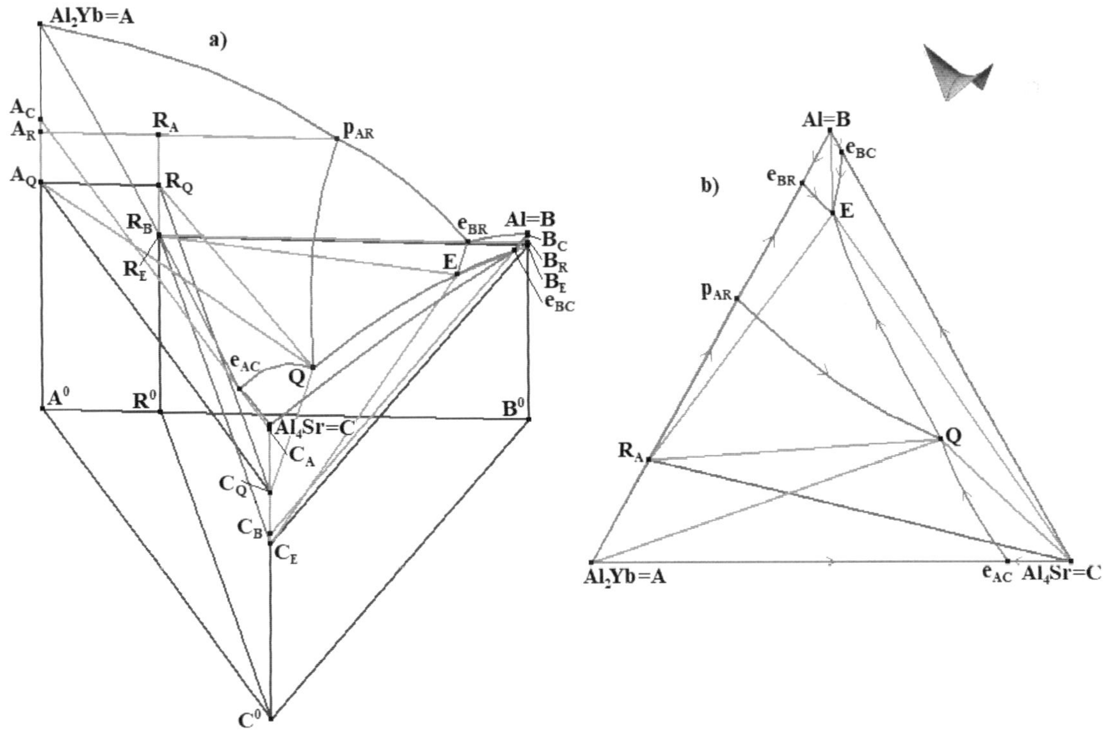

Figure 2. 3D computer model (a) and x-y projection (b) of T-x-y diagram Al_2Yb-Al-Al_4Sr.

4. Calculation of mass balances and analysis of concentration fields

At the projecting of all elements of T-x-y diagram into concentration simplex, it is divided on two-dimensional, one-dimensional and zero-dimensional concentration fields. The particularities of occurring processes in this fields are studied using the diagrams of vertical mass balance, which show an increase or decrease of phase portion for each phase region. [11-12]. They also make it possible to compare the processes taking place in different concentration fields and qualitatively show the concentrations fields with different by phase schemes, but with similar microstructure [11-12].

So the projection of studied subsystem is divided into 14 two-dimensional fields (1-14), 18 one-dimensional ($e_{AC}Q_c$, Q_cQ, $p_{AR}E_R$, E_RQ, QE, $e_{BR}E$, $e_{BC}E$, Q_AA, Q_AQ, CQ_c, CQ, BE, CE, RQ_A, Q_AQ_c, RQ, E_RR, E_RE) and 5 zero-dimensional (Q, E, Q_e, Q_A, E_R) concentration fields (Fig. 3a). Using diagrams of vertical mass balance, it was revealed that five concentration fields coincide by the crystallization stages and microstructural sets with neighboring fields of higher dimension $E_R \in E_RE$, $Q \in QE$, $p_{AR}E_R \in 6$, $E_RQ \in 5$, $CQ \in 10$. One more one-dimensional field Q_eQ coincides with field 12 by the microstructural set, but differs in the crystallization stages [13-14].

Let's consider in more detail the diagrams of vertical mass balance calculated for fields 12 and Q_cQ.

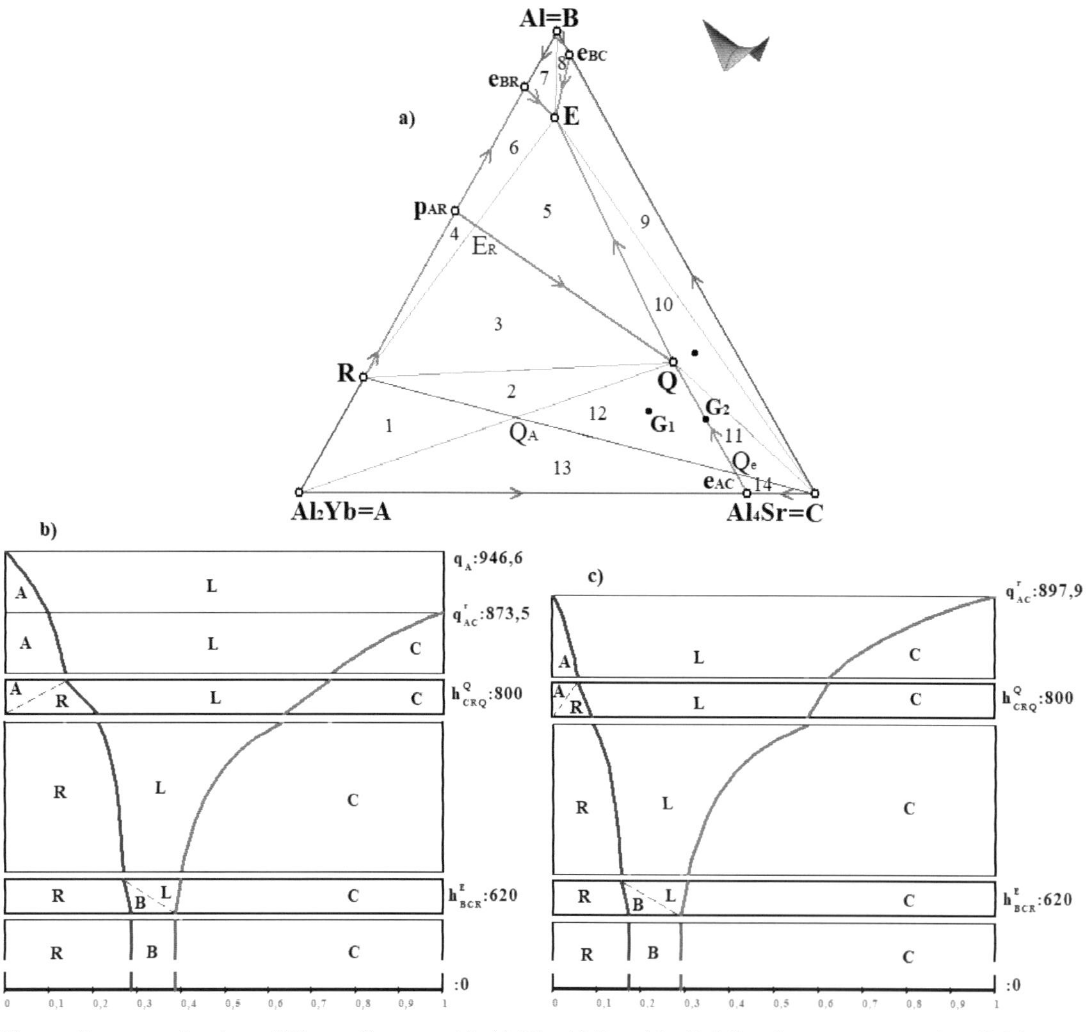

Figure 3. x-y projection of T-x-y diagram Al-Al_2Yb-Al_4Sr with dividing into concentration fields (a); diagrams of vertical mass balance for mass centers $G_1 \in 12$ (b) and $G_2 \in Q\text{-}Q_e$ (c).

The mass center G_1 (0.213; 0.173; 0.386) giving in field 12 (Fig. 3b) intersects the liquidus surface q_A at 946.6°C and falls in two-phase region L+A, where the portion of phase A increases and the portion L decreases, which corresponds to primary crystallization $L^1 \rightarrow A^1$. Then it crosses the ruled surface q^r_{AC} on the boundary of three-phase region L+A+C, in which, along with an increase of portion of phase A, the phase C appears and grows, while the content of L decreases, i.e. monovariant eutectic crystallization occurs $L^e \rightarrow A^e + C^e$. Further, on the horizontal complex at the temperature of the invariant quasiperitectic point Q, a four-phase regrouping of the masses occurs $L_Q + A \rightarrow C_Q + R_Q$ at 800°C, as a result of which the crystals A^1 and A^e are completely replaced by crystals R. Then the mass center falls into the three-phase region L+C+R, where the growth of phases C and R corresponds to the postperitectic secondary eutectic reaction $L^{ep} \rightarrow C^{ep} + R^{ep}$. On the horizontal complex at the temperature of ternary eutectic E, the invariant eutectic reaction $L_E \rightarrow B_E + C_E + R_E$ at 620°C ends with the deficiency of melt and crystals B, C and R remain below. Thus, the field 12 is characterized by the following set of microstructural components: C^e, C_Q, R_Q, C^{ep}, R^{ep}, B_E, C_E, R_E. Crystals A^1 and A^e are not included in this set, because completely consumed at invariant quasiperitectic reaction.

The diagram of vertical mass balance for center of mass G_2 (0.13; 0.161; 0.709) giving for neighboring field Q_eQ does not include the primary crystallization reaction (Fig. 3c), since this one-dimensional field is the part of a monovariant liquidus line $e_{AC}Q$. The mass center for this field will immediately fall into the three-phase region L+A+C. Otherwise, the crystallization stages coincide with field 12. Therefore, fields 12 and Q_eQ differ in the crystallization stages, but coincide with set of microstructural components.

The analysis of all concentration fields for the subsystem Al_2Yb-Al-Al_4Sr was carried out in a similar way.

5. Summary

Frederick N. Rhines gave some useful advices [3] to master the language of phase diagrams: "Each phase region is modeled separately in clay of different colour, and the various pieces are fitted into the completed diagrams. When finished, the model may be cut horizontally or vertically to obtain the isotherms and isoplets. Another very satisfactory method of making a model of a ternary diagram is by the use of wires to represent all the lines in the diagram. Cardboard may be used and coloured threads strung horizontally around the wires bounding the three edges of each three-phase region will serve to produce tie-triangles. This kind of models cannot be sectioned, of course, but it has the advantage, that the structure can be "seen through" and each phase region observed".

In our days, thanks to the computer models of T-x-y diagrams, this task (to master the language of phase diagrams) may be solved by means of a 3D printer. After the prototyping, the phase regions will be used to puzzle the so-called exploded phase diagrams.

One more application of the T-x-y diagrams computer models is connected with the opinion of William Hume-Rothery [15]. He has estimated, that experimental building of a T-x-y diagram with medium-complexity will require the collaboration of two experienced researchers over a period of 5-10 years.

Computer models of the phase diagrams will help to do it in more short time. Thanks to the information on material balances, it is possible to generate the DTA spectrum prototype for any composition. The peak height of the DTA signal represents the phase with the maximal increase or decrease of mass portion in the corresponding temperature interval. Not gear but rectangular peak corresponds to the invariant regrouping of masses on the horizontal plane. The melt presence is showed by the elevated level of electrical conductivity. For generation of X-ray spectra the training software has been fitted by the data base with characteristics of initial components A, B, C and corresponding binary and ternary compounds (with their allotropies).

Main destination of PD computer model – to solve practical problems, beginning from correct and compact representation of the experimental information and finishing the calculation of formed in the system microconstituents of the phase assemblages. Diagram of vertical mass balances give a possibility to follow the changing of phase's mass fraction for any composition within the temperature interval. Horizontal mass balances help to show the phases portions on the isopleth at the given temperature.

When the phase regions are projecting on the concentration simplex, the latter is dividing into the concentration fields with the individual sets of phase reactions and appropriate microstructures. Then the fields, which belong to the 3-phase regions, are dividing additionally by the surfaces of 2-phase reaction into the fragments with the different dynamics of phase masses increment. One more reason for the microstructures variety is the competition of crystals with different dispersity, when a field of invariant reaction is divided into the fragments with the tiny eutectic crystals, with more large primary crystals and with the both types of these crystals.

Thus a kind of the training software will allow to create the PD computer model and then to construct vertical and horizontal sections with the decoding of intersected surfaces and phase regions, to calculate the mass balances, to generate the spectra of thermal and X-ray phase analyses for given composition. A user will assume the PD topology by means of the binary systems information, and then will determine all geometrical elements by means of minimal number of experiments.

Acknowledgement

This work has been performed under the program of fundamental research SB RAS (project 0336-2019-0008) and was partially supported by the Russian Foundation for Basic Research project № 20-21-00056.

References

[1] Schmid-Fetzer R 2014 *J. Phase Equilibria and Diffusion* **35** 735–760.
[2] Prince A 1966 *Alloy Phase Equilibria,* Elsevier: Amsterdam-London-N.Y.
[3] Rhines F N 1960 *Diagrams of Phase Equilibria in Metallurgy*. Moscow, 1960. (In Russian)
[4] *State Diagrams of Binary Metal Systems.* Vol. 2. 1997 ed. N P Lyakishev, Mashinostroyenie, Moscow (In Russian).
[5] Rakhimov F K 2015 Phase diagrams and thermodynamic properties of europium and ytterbium alloys *Dissertation* (In Russian).
[6] Lutsyk V I and Vorob'eva V P 2008 *Z. Naturforsch. A.* **63a** 513-518.
[7] Lutsyk V I, Vorob'eva V P and Sumkina O G 2012 *Crystallography Reports* **57** 115-125.
[8] Lutsyk V I and Zelenaya A E 2018 *Russ. J. Inorgan. Chem.* **63** 966-973.
[9] Lutsyk V I and Vorob'eva V P 2016 *Russ. J. Inorgan. Chem.* **61** 858-866.
[10] Lutsyk V I, Zelenaya A E and Zyryanov A M 2008 *Journal of Materials, Methods & Technologies, International Science Publications* **2** 176-184.
[11] Lutsyk V and Zelenaya A 2017 *Journal of Physics: Conference Series* **790** 012020.
[12] Lutsyk V I and Zelenaya A E 2018 *Russ. J. Inorgan. Chem.* **63** 1087-1091.
[13] Lutsyk V I and Vorob'eva V P 1995 *Russ. J. Neorg. Chem.* **40** 1697-1703 (in Russian).
[14] Lutsyk V I and Vorob'eva V P 1997 *Russ. J. Phys. Chem. A.* **71** 399-402 (in Russian).
[15] Hume-Rothery W, Christian J W and Pearson W B 1952 *Metallurgical equilibrium diagrams*, London: The Institute of Physics.

Porous silicon obtained by the metallic thermal reduction of high-purity silica gel

A Nomoev[1,2,*], V Lygdenov[1,2], V Lygdenov[1], V Mankhirov[1], E Khartaeva[1], V Syzrantsev[1]

[1]Institute of Material Physics of Siberian Branch of Russian Academy of Science, 6 Sakhyanovoi Street, Ulan-Ude, 670047, Russia
[2]Department of Physics and Engineering, Buryat State University, 24a Smolin Street, Ulan-Ude, 670000, Russia

E-mail: nomoevav@mail.ru

Abstract. The synthesis and structure of porous silicon obtained through metallothermic reduction of high-purity silica gel is described. The obtained material was a spongy formation of porous amorphous-crystalline silicon. The described reaction is cost effective, rapid, and a simple method to obtain porous silicon particles.

1. Introduction

Porous silicon is a material with unique properties such as high adsorptive activity, catalytic activity, and optical properties [1,2]. The developed surface area can dramatically increase the quantum yield of electrons from electrodes made of porous silicon in electric batteries as well as the possibility of using porous silicon as a cathode with a low electron work function [3,4]. Industrial methods for the preparation of silicon mainly consist of the carbothermic reduction of silicon dioxide [5] and purification by direct chlorination of silicon [6].

Porous silicon nanomaterials are increasingly attractive for biomedical applications, for example, as drug delivery vehicle [7] due to their promising properties such as simple and feasible fabrication procedures, tunable morphology, versatile surface modification routes, biocompatibility and biodegradability [8].

In general, high purity silicon can be produced in two steps [9]: First, a metallothermic reduction material with a high content of silica and magnesium or aluminiumis chemically treated with hydrochloric and hydrofluoric acid. The by-products of the reactions are then removed.This process is summarized by the following reaction conducted in an inert argon atmosphere:

$$SiO_{2(s)} + 4Mg_{(s)} \rightarrow Si_{(s)} + Mg_2Si_{(s)} + 2MgO_{(s)} \quad (1)$$

To maximize the amount of silica and reduce the formation of the silicide, the optimal ratio for the reacting components is a 4.5:5.5 weight ration of Mg: SiO_2. [10]. The metallothermal reaction is highly exothermic and thus an additional material such as NaCl should be added to absorb excess heat [11]. In order to remove the Magnesium silicide and oxide from the product, HCl and HF are used to cause the following reactions:

$$Mg_2Si_{(s)} + 2HCl_{(aq)} + H_2O_{(l)} \rightarrow MgCl_{2(aq)} + SiH_{4(g)} + MgO_{(s)} \quad (2)$$

$$MgO_{(s)} + 2HCl_{(aq)}O \rightarrow MgCl_{2(aq)} + H_2O_{(l)} \tag{3}$$

$$SiH_{4(g)} + 2O_{2(g)} \rightarrow SiO_{2(s)} + 2H_2O \tag{4}$$

$$SiO_{2(s)} + 4HF_{(aq)} \rightarrow SiF_{4(g)} + 2H_2O_{(l)} \tag{5}$$

In the process of adding HCl the powders obtained from the thermochemical reactions (1), the silicon powder settles to the bottom of the reaction vessel and magnesium chloride goes into solution and can beremoved by filtration with water. The silane (SiH4) gas combines with oxygen and ignites spontaneously to form silicon dioxide and water by reaction (4). This new mixture is then treated with concentrated hydrofluoric acid, HF,to remove the remaining silicon dioxide by producing silicon tetrafluoride (SiF_4) and water (reaction (5)).
Usually, porous silicon is produced on the surface of monocrystalline silicon for the needs of electronic devices (photoluminescence) [12]. However, powdered porous silicon is not usually produced. The present work aims to address this issue and give a way to simply obtain chemically pure porous silicon powders. The synthesis of porous silicon with a high specific surface area and a nanoscale structure from high purity silica gel is presented. These powders have potential to be used as carrier particles in medicine and biotechnology [13]. In addition structure makes it a prospective material for catalysts, microelectronics, and conductive materials.

2. Experimental section

High-purity silica gel for chromatography (PURASIL TM 60Å) with grains of 63-200 micronswere used as the starting silicon source. The reducing agent was chemically pure magnesium powder (Russian standard 6001-79).For comparison,instead of pure magnesium powder, waste magnesium from aircraft production was also tested (ML10, ML12, Ml19 from Russian standard 2856-79 and wrought-MA8, MA8 from Russian standard 14957-76) which contain alloying elements that are readily soluble in inorganic acids. Synthesis was carried out in a steel reaction apparatus as shown in Figure 1. Argon was used as a protective inert gas at high-temperature synthesis of silicon in the working space of the reactor (Fig. 1.).

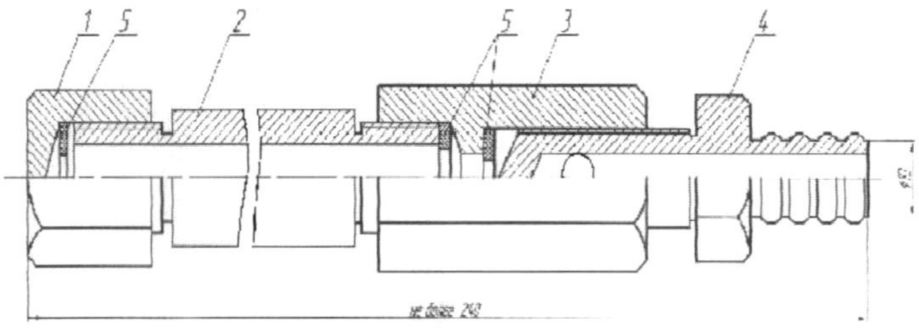

Figure 1. Schematic of the steel reaction apparatus.
1 -End cap, 2–reaction chamber, 3 -sealing cap, 4 -nozzle for feeding in argon gas and for locking the working space of the reaction chamber, 5 –copper sealing washer.

High-purity silica was thoroughly mixed with magnesium in a ratio of 5.5: 4.5 by weight. To absorb heat from the exothermic reduction reaction and to preventing sintering of the produced particles, NaCl was added in an amount of 1: 1 and 1: 2 by weightrelative to the silica-magnesium powder. In previous works, ratios of 1:10 were used [11]. The powders were placed in the reaction apparatus, filled with argon, then placed in a preheated oven at 700 °C.

The reaction time was 1 to 2 hours; depending on the quantity of NaCl. Such a temperature is required for evaporation of the active magnesium reducing agent.
The NaCl was washed out from the reduced product containing silica, magnesium silicate, and magnesium oxide by filtration with distilled water. The powders were then dried and treated with concentrated hydrochloric acid followed by thorough washing in distilled water. Finally a treatment with hydrofluoric acid was performed followed by filtration and washing.
The final powders were analysed using energy dispersive X-ray fluorescence analysis (EDX),X-ray diffraction (XRD), surface area analysis, Fourier transform infrared spectrometry (FTIR), and scanning electron microscopy.EDX and SEM were performed on aSEM JEOL 6510 LV (accelerating voltage 30 kV, a resolution of 3 nm). XRD was performed on a D8 ADVANCE Bruker, Germany (CuKα radiation with a wavelength of 0.15418 nm). Determination of the specific surface area was performed on a "Sorbie-M", Novosibirsk, Russia. FTIR was performed on a Shimadzu-8300 at a resolution of 4 cm-1 with 200 scans.

3. Results and Disussion

XRD analysis of the synthesized powders before any post-process treatments is shown in Figure 2. The analysis revealed the presence of crystalline Si, MgO, $MgSiO_3$, and an amorphous component amounting to about 56.2 wt% of the sample. It was interesting and unexpected that no Mg_2Si was formed and instead the magnesium silicate $MgSiO_3$ was formed.It appeared that when the powder was treated with HCl no combustion was observed and thus it is likely that the reaction

$$MgSiO_{3(s)} + HCl_{(aq)} \rightarrow MgCl_{2(aq)} + SiO_{2(s)} + H_2O_l \quad (6)$$

occurred as SiH_4 did not form through equation (2). Thus, the formation of $MgSiO_3$ instead of Mg_2Si makes the synthesis safer.

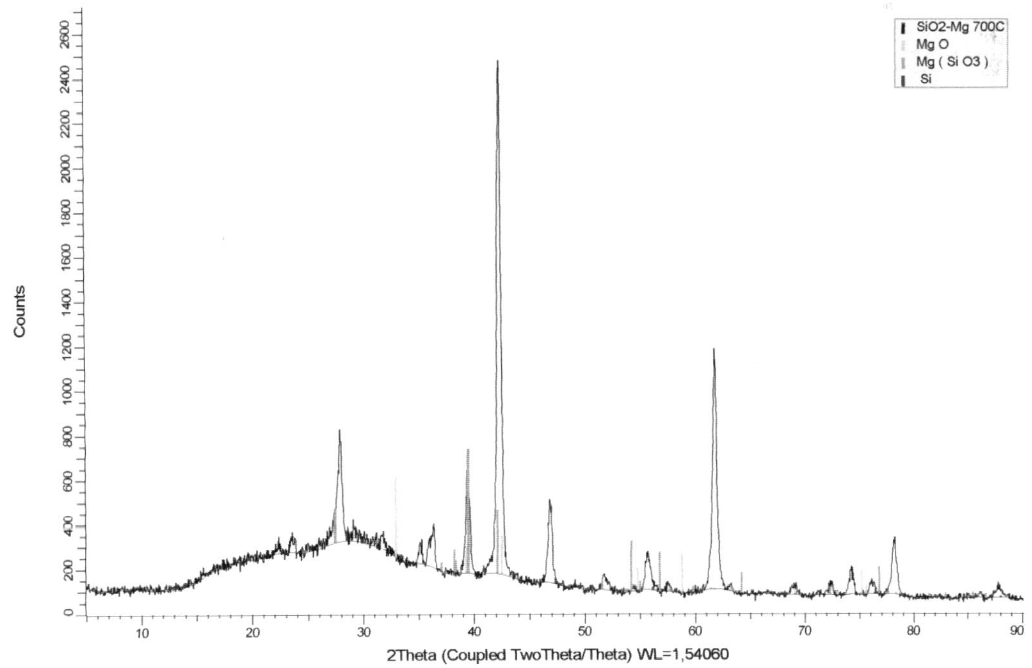

Fig.2. X-ray diffraction analysis of the synthesized porous silicon powder after HCl treatment

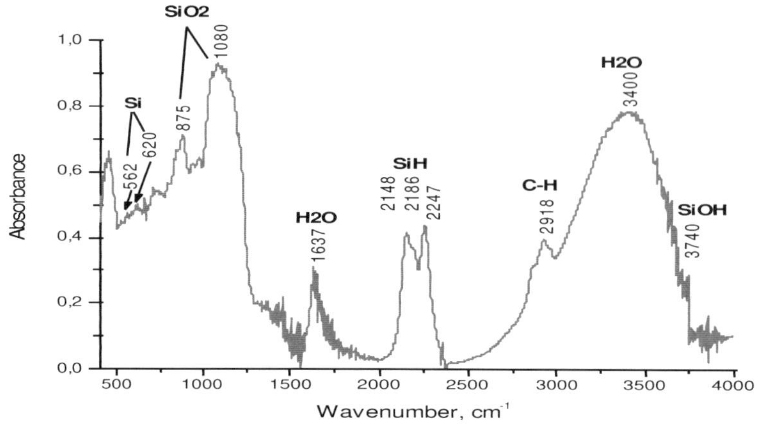

Figure 3. IR spectrum of the silicon powder obtained from the silica gel before treatment.

The IR spectrum of the synthesized powder before any treatments is shown in Figure 3 to investigate possible components of the amorphous phase. Hydrides of silicon were found (possibly with magnesium) around the 2100 cm-1 band. Silica was present in the sample as evidenced by the bands at 480, 875, and 1080 cm-1. Silica was also identified by the SiOH group at 3742 cm-1. The amount of silica is low, since the intensity of the 3742 cm-1 bands are significantly lower compared with the hydrides and it is well-known that the extinction coefficients of SiH bands are significantly lower compared to SiOH bands.

To determine the proportion of silicon oxide and silicon metal in a sample was not possible as the bands used to identify silicon at 562 and 620 cm-1 likely to belong to the defect structure of silicon metal and thedefects for different silicon samples can vary greatly. However, if this is not taken into account, it can be estimated that silica makes up only 1-2wt% of the amorphous component. It follows that most of the synthesized was in the amorphous state.

Figure 4. SEM images of the final silicon powder after treatment with HCl and HF which were synthesized with a)Mg powdersand b) magnesium alloy chips.

SEM images of synthesized and treated silicon powders are shown in Figure 4; from two different magnesium sources. It is shown that the powders have a spongy structure and the final appearance did not change significantly between them. Other analyses did not show significant differences and so waste magnesium alloys can also be used to reduce the cost of the synthesis.

The high porosity and open structure of the silca gel used in this synthesis reduced the synthesis time required to produce porous silicon powders compared to previous reports such as in [9] which used sand as the starting material. When the reactant;NaCl ratio was 1:1, the specific surface area for the final powders was 200 m^2/g, and when the content was 1:2, the specific surface area was 160 m^2/g. Thus, increasing the salt content in the mixture lead to a decrease in the specific surface of the synthesized silica. However, if no NaCl was used, the reaction did not proceed at 700 °C and at a higher temparature of 1000 °C spheres of silicon melted out of the stucture and did not show any porsity. The specific surface of the synthesized silica was lower compared to the specific surface of the silica gel (250-350 m^2/g) [14]. It is expected that the main cause of porosity for the resultant silicon is a retention of the structure of the silica gel which is thermally stable to 900 °C. Some silicon may have been released from the starting structure during the synthesis and treatment processes and later re-condensed onto the silica/silicon and NaCl – resulting in some decrease of the surface area. With higher NaCl content, the released silicon may have condensed onto NaCl (m.p. ~ 800 °C) more than the porous silica/silicon and formed denser structures which reduced the overall surface area of the sample.

TEM analysis results show that silicon is stood out in the form of a porous, the pore size of less than 10 nm and above (Figure 5). Moreover, silicon is linked together in the form of crossed fibers with a diameter of about 10 nm. Such a structure is particularly important for making electrodes with high surface area, which enhances electron emission.

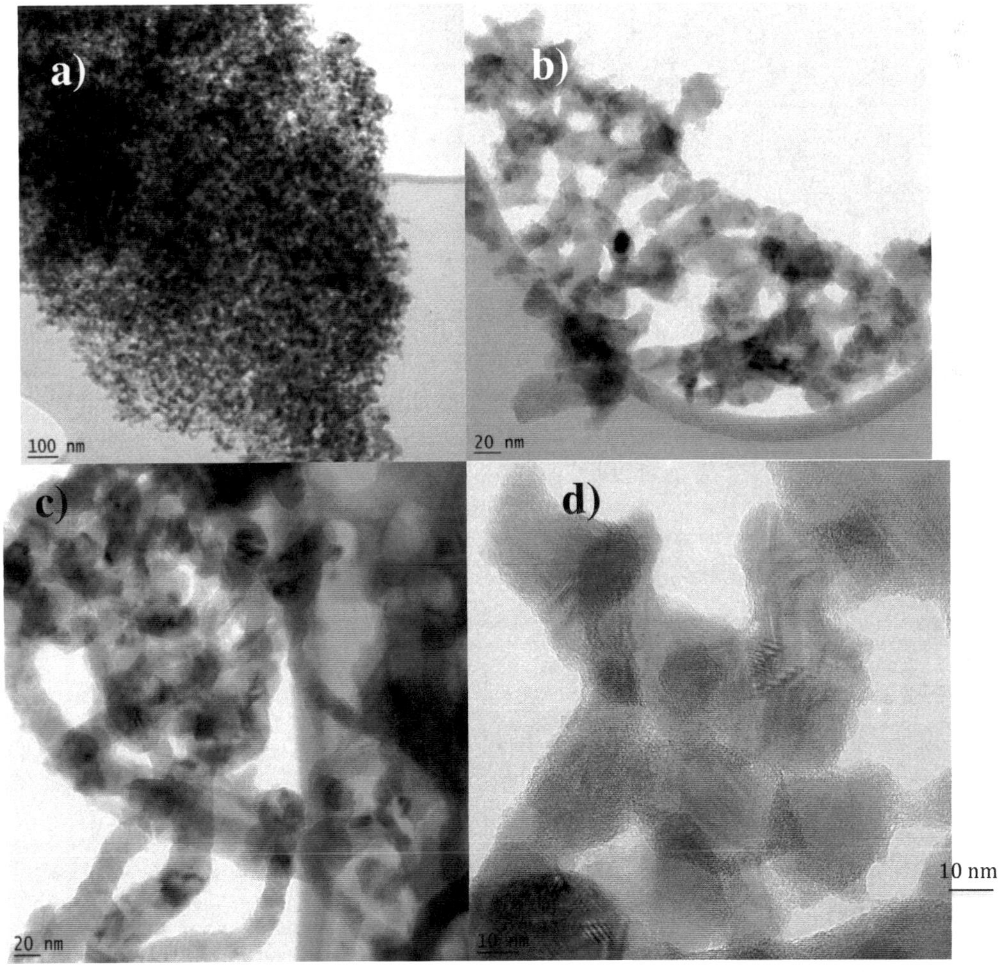

Figure.5. TEM images of the final silicon powder after treatment with HCl and HF which were synthesized withMg powders: a – x12000, (b, c) - x50000,d - x150000.

Elemental analysis EDX (Figure 6) showedonly silicon and oxygen. Some oxidation may have occurred after removal of the reacted powders from the argon environment and treatment process. Thus oxidation may be hindered if the treatments are performed in a reducing (CO, H_2) or inert (N_2, Ag) atmosphere. Additioanly, it was observed that less oxidation was present in the samples produced from waste magnesium trimmings. Further studies will be required to investigate the effects of the alloying elements in the waste magnesium trimmings on the synthesis process.

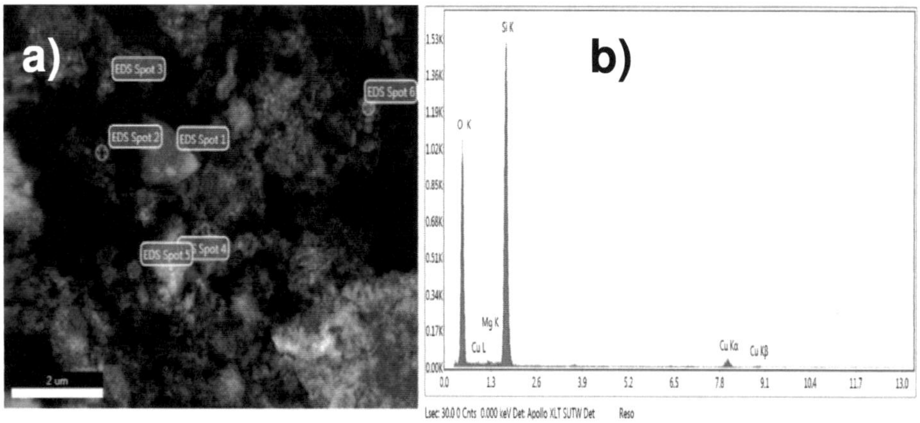

Figure 6. a) SEM image of obtained porous silicon.
b) EDX analysis of the porous silicon powder..

4. Conclusion

The synthesis process of porous silicon powders were demonstrated. The synthesis utilizes silica gel heated to 700 °C with magnesium and NaCl. It was shown that waste magnesium trimmings can be used instead of pure magnesium powders which can further reduce the synthesis costs. Further, the synthesis times were short at about 1-2 hours. Post treatment processes involve washing the synthesized powders with HCl and HF. The resultant silicon had a spongy, mainly amorphous structure and had specific surface areas between 160-200 m2/g.

5. Aknowlegments

This work was financially supported by the project No. 0336-2019-0008 of the Russian Ministry of Education and Science and RFBR grant No.18-42-030004.

6. References

[1] Chaudhuri R G and Paria S 2012 *Chem. Rev.*, **112**, P. 2373–2433
[2] Srdić V V, Mojić B, Nikolić M and Ognjanović S 2013 *Processing and Application of Ceramics* **7** 45–62
[3] Beale M I J, Benjamin J D, Uren M J, Chew N G and Cullis A G 1985 *J. Cryst.Growth.* V 73 **2** 622-636
[4] Chernyakhovsky L V 2010 *Russian Patent* **2383493** C1
[5] Matheson A 2010 *US patent* **2010107275** A
[6] Fengs Z C and Tsu R 1994 *World Scientific* 465
[7] Zhang D-X, Esser L, Vasani R B, Thissen H and Voelcker N H 2019 *Nanomedicine* V 14 **24** 3213-3230
[8] Alhmound H, Delatat B, Elnathan R, Cifuentes-Rius A, Chaix A, Rogers M, Durand J and Voelcker N H 2015 *Adv. Funct. Mater.* **25** 1137- 1145

[9] Smith R L and Collins S D 1992 *J. Appl. Phys.* **71** R1
[10] Razmyslov V I 1995 *Russian patent* **2036143** C1
[11] Favors Z J, Wang W, Bay H H, George A and Ozkan C 2014 *Sci.Rep.*4
[12] Vial J C and Derrien J 1995. *Springer* 556
[13] Hughes G A 2005 *Nanomed-Nanotechnol.* **1** 22–30
[14] Neimark I E and Shane-Fine R J 1973 *NaukovaDumka* 27-29

IOP Conf. Series: Materials Science and Engineering 1000 (2020) 012010 doi:10.1088/1757-899X/1000/1/012010

Terahertz photoconductive antenna with embedded electrodes: simulation and experiment

S Nomoev[1, *], I Vasilevskii[1], A Klochkov[1], A Vinichenko[1] and E Khartaeva[2]

[1]National Research Nuclear University MEPhI (Moscow Engineering Physics Institute), Kashirskoe highway 31, Moscow, 115409, Russia
[2]Institute of Material Physics of Siberian Branch of Russian Academy of Science, 6 Sakhyanovoi Street, Ulan-Ude, 670047, Russia

E-mail: serganom@gmail.com

Abstract. The paper presents the results of numerical simulation of a terahertz (THz) photoconductive antenna with embedded electrodes by the finite element method. The simulation results indicate the fact that the proposed THz antenna has a higher photocurrent than the conventional photoconductive antenna with conventional electrode contacts. Higher THz power can potentially be obtained using the proposed photoconductive antenna with embedded electrodes. The simulation results show that the electric field strength at the surface is higher for conventional PCA, however, the PCA depth with embedded contacts has a higher electric field strength. The simulation results show that the increase in the photocurrent is directly proportional to the thickness of the embedded contacts. The results of the performed experiments are consistent with the conclusions of the simulation.

1. Introduction

The development of efficient compact antennas for transmitting and receiving terahertz waves is urgent. Recently, works have been published, some of which are considered below, devoted to the creation of photoconductive antennas (PCA) that increase their efficiency due to a change in the mesa structure of contacts, as well as modeling of the electrophysical parameters of THz (PCA) depending on the type of contacts.

InGaAs / InAlAs photoconductive antennas emitting radiation at a wavelength of 1.5 μm have been improved by mesa-etching of the conducting layers [1]. Electrical side contacts were applied to the multilayer structure, and the layers are removed in areas that are only prone to parasitic dark currents. Consequently, photocurrents are increased 5 times, and parasitic dark currents are suppressed. The radiated terahertz power is increased by more than 5 times compared to conventional planar antennas, the sensitivity of the receiver is increased by 11 times. Overall, the mesastructured antenna outperforms previous planar systems by a factor of 27.5 in output signal amplitude.

A metal-semiconductor-metal (MSM) photodetector based on low-temperature GaAs with doped (that is, ohmic type) contacts have been manufactured and characterized [2]. Burn-in contacts optimize the distribution of the electric field within the structure of the photodetector, which leads to an increase in the sensitivity of the devices by up to 200% compared to conventional MSM detectors with standard Schottky-type metalized metallization made on an identical material.

Thus, in [3], a simulation of the PCA was carried out, where the substrate used is a low-temperature gallium arsenide LT-GaAs, with a hole mobility of 400 $cm^2/V * s$ and electron mobility of 8500 cm^2/V

* s. A linear dependence of the strength of the photocurrent excited by a picosecond laser pulse, PCA, on the bias voltage is observed. Comparative analysis of the photocurrent of an antenna with two-fold lower mobility of charge carriers showed a directly proportional dependence of the calculated theoretical values of the photocurrent on the bias voltage and a higher photocurrent for PCA with higher values of the mobility of charge carriers.

A photoconductive terahertz antenna based on a distributed Bragg reflector, nanoplasmonic gratings, and recessed electrodes was proposed in [4]. Using the finite element method and two-wave modeling, the influence of geometric parameters on the nonstationary photocurrent of the proposed photoconductive antenna is investigated. The nanoplasmonic structure reduces the reflection of laser light by up to 1.5% from the surface of low-temperature gallium arsenide, compared to 29% for a conventional photoconductive antenna. According to the results, a distributed Bragg reflector in combination with a nanoplasmonic grating and recessed electrodes leads to an increase in the photocurrent peak by 5265% compared to a conventional terahertz photoconductive antenna.

A THz PCA with a plasmon grating, a metal electrode height h = 100 nm, and an aspect ratio h / p = 0.5 (p is the period of the plasmon grating) was proposed in [5]. Higher power of generated THz radiation in a plasmon PCA, which is two orders of magnitude higher than in an equivalent PCA without a plasmon grating, has been experimentally demonstrated. This is due to a sharp increase in the electric field near the metal-semiconductor contact in the plasmon lattice, in connection with which a larger number of photoexcited current carriers reach the antenna contacts and contribute to the generation of THz radiation.

Thus, the efficiency of the PCA antenna can be increased in two ways: the first is the use of embedded contacts, the second is the use of nanoplasmonic gratings in the interelectrode gap.

The work aims to simulate the PCA of an antenna with embedded contacts, create a PCA with embedded electrodes, and compare the experimental data with the simulation results.

2. Simulation and experimental methods

2.1. Simulation

The model consists of two related studies - stages; first, the optical response is calculated in the frequency domain using the electromagnetic wave equation (1).

$$\nabla \times \mu_r^{-1}(\nabla \times \mathbf{E}) - \mathrm{k}_0^2(\varepsilon_r - \frac{j\sigma}{\omega\varepsilon_0})\mathbf{E} = 0 \qquad (1)$$

The laser source is modeled as a focused monochromatic Gaussian beam at a wavelength of 800 nm. From here it is possible to calculate the Poynting vector and, as such, the photon absorption flux inside the PCA substrate. The advantage of including this step in the model is that the photon absorption flux in the PCA substrate is not uniform. Although uniform absorption approximations have been shown to work for traditional THz PCA, it is well known that plasmonic photoconductive devices have an extremely inhomogeneous distribution of the electric field, with most of the electromagnetic energy highly concentrated near the plasmonic structure [2].

The second study - the stage determines the dynamics of carriers due to the applied bias V and the optical generation rate of charge carriers $U_{n,p}$. The peak generation of carriers is taken from the photon absorption flux calculated in the first stage, an optical study, and is approximated by a Gaussian time dependence. The temporal pulse width corresponds to the width of the femtosecond laser pulse (full width at the half-amplitude level is ≈ 100 fs). The equations governing the dynamics of these charge carriers in the semiconductor layer are the Poisson equation (2) and the carrier transport equations for electrons (3) and holes (4):

$$\nabla \cdot (\varepsilon_r \nabla \mathbf{V}) = q(\mathbf{n} - \mathbf{p} + N_A - N_D) \qquad (2)$$

$$\frac{\partial \mathbf{n}}{\partial t} = -\frac{1}{q}\nabla J_n - U_n \tag{3}$$

$$\frac{\partial \mathbf{p}}{\partial t} = \frac{1}{q}\nabla \mathbf{J}_p - U_p \tag{4}$$

Here n and p are the concentration of electrons and holes, Na is the concentration of acceptors, Np is donors, Un = Rn + Gn, where Rn is the rate of electron recombination, Gn is the rate of electron generation, Up = Rp + Gp, where Rp is the rate recombination of holes, Gp is the rate of generation of holes, Jn, p is the current density of electrons/holes. The current density is a function of the electron/hole concentration, material properties, and the applied potential V. Equations (2-4) are then directly related through V, n, and p, while equation (1) defines the spatial distribution Un, p which depends only on time.

To check the correctness of the calculations, a simulation of a static electric field was carried out for two copper plates with a distance between them of 100 μm with a voltage between the plates of 2 V, shown in the figure. It can be seen that the lines of force of the electric field, marked with white arrows (the magnitude of the electric field depends on the thickness of these arrows) and the potential distribution correspond to the classical distribution of the field between two metal plates. Modeling was carried out by solving Maxwell's equations. Under static conditions, the electric potential V is determined by the ratio:

$$\mathbf{E} = -\Delta V \tag{5}$$

Combining this equation with the defining relationship:

$$\mathbf{D} = \varepsilon_0 \mathbf{E} + \mathbf{P} \tag{6}$$

between the electric displacement D and the electric field E, one can represent Gauss's law as the following equation:

$$-\nabla \cdot (\varepsilon_0 \nabla V - \mathbf{P}) = \rho \tag{7}$$

In this equation, the physical constant ε_0 (SI unit: F/m) is the dielectric constant of vacuum, P (SI unit: C/m^2) is the electric polarization vector, and ρ (SI unit: C/m^3) is the space charge density ... This equation describes the electrostatic field in dielectric materials.

The proposed model was used to calculate the photocurrent as a function of the applied bias voltage. In fig. 2 shows a schematic diagram of photoconductive antenna structures and their corresponding dimensions for a conventional and proposed photoconductive antenna with fired electrodes, respectively. This is a model geometry, which differs in size from the real topology of contacts (The gap between the electrodes is 200 μm, the thickness is 100 μm, the height is 900 nm. In the model, the gap between the electrodes is 1 μm, the thickness is 2.2 μm and the height is 80 nm), due to for problems in modeling at real dimensions (The Gaussian beam is blurred for such dimensions. The interelectrode gap becomes more than 2 orders of magnitude of the laser radiation wavelength). LT-GaAs was used as the substrate material, while gold was used as the electrode material. The simulated values of the electric field and photocurrent are constructed using a relatively constructed coordinate system, as shown in Fig. 2.

Simulation of the static electric field created by the bias voltage and applied to the electrodes for two antennas with different types of electrodes has been carried out. In Fig. 1 a, b, white lines show the lines of force of the static electric field. The thickness of these lines is directly proportional to the strength of the electric field. On the surface of the LT-GaAs semiconductor between the electrodes in an antenna with burn-in contacts (Fig. 1a), the concentration of thick lines is higher compared to conventional electrodes (Fig. 1b). Thus, the fired electrodes increase the area of the static electric field on the surface of the semiconductor between the electrodes in the antenna, created by the bias voltage in comparison with conventional electrodes.

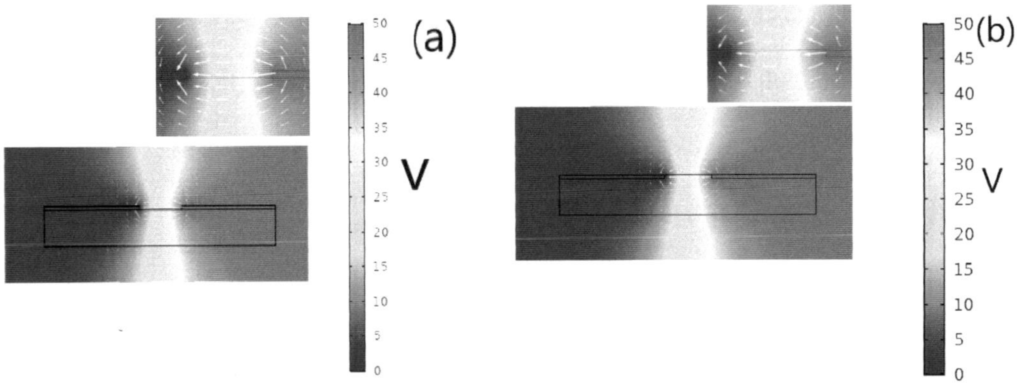

Figure 1. The lines of force of the static electric field and potential in the LT-GaAs antenna with conventional electrodes - a, with embedded electrodes - b.

A higher dc field in the near-surface layer of LT-GaAs leads to greater capture of photocarriers by electrodes, which increases the output photocurrent of the PCA.

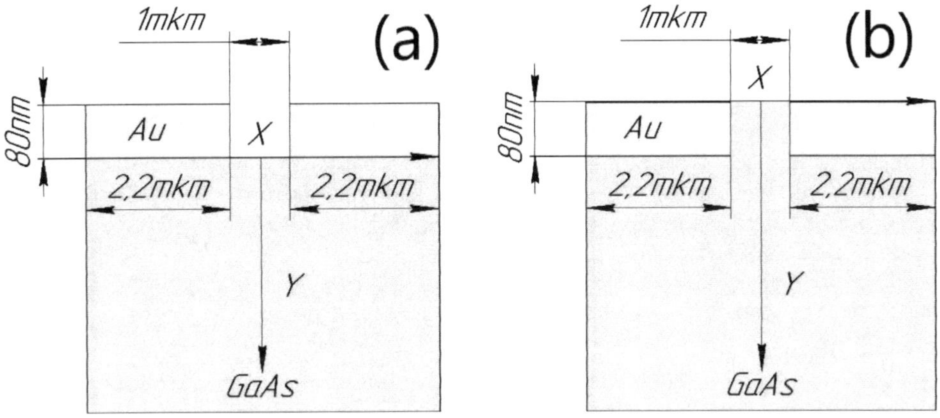

Figure 2. Schematic diagram of the structure of the simulated photoconductive antenna; (a) - conventional model, (b) - with embedded contacts.

Obviously, recessed electrodes can better collect photogenerated carriers, resulting in significantly increased sensitivity. Simply put, in devices with recessed electrodes, incident photons are absorbed in a larger volume of LT-GaAs, where a relatively uniform high electric field provides the most efficient collection of carriers.

Unlike microwave antennas, a photoconductive antenna in the millimeter range (wavelength range 1 mm - 30 μm) is powered by two different sources, namely from DC voltage and optical pumping. The finite element method, which combines the Maxwell wave equation with the Poisson drift-diffusion equations, allows one to calculate the density distribution of photogenerated carriers generated by a pump laser. Laser radiation characteristics: laser wavelength 800 nm, paraxial Gaussian beam with a beam radius of 800 nm. These data are necessary to calculate the distribution of the electric field inside the photoconductive device, photocurrent when a bias voltage is applied to the electrode.

2.2. Making experimental samples and measuring dark current

The PCA under study (samples 1 - 4) were fabricated on an epitaxial structure with an LT-GaAs film 1.2 μm thick. The LT-GaAs film was obtained by molecular beam epitaxy (MBE, Riber Compact 21

setup) at a temperature of 230 ° C and a pressure ratio of As_4 molecules and Ga atoms equal to 19. After growth in the MBE chamber, the structure was annealed in an As flow at 580 ° C.
Flag-type PCA with metallization (Ni / Ge / Au / Ni / Au) was manufactured by standard methods of planar technology. The antenna geometry is shown in Figure 2. The width of the photoconductive gap between the PCA electrodes was 200 μm. For PCA №1, the contact embedded procedure was not performed, and for PCA №2-№4, a 2-minute contact firing was performed at a temperature of 410 ° C. The thickness of the metalized contacts is 900 nm. The procedure for fabricating the PCA (samples 2 - 4) - growth, annealing of the LT-GaAs crystal, and embedding of contacts are identical.

Figure 3. The geometry of electrodes manufactured by PCA

The volt-ampere characteristics (I - V) of the antennas were measured using an EP6 handheld probe station (Cascade Microtech, USA) and an Agilent B1500A parametric analyzer (Agilent Technologies, USA). The bias voltage was of different polarities and varied from -50 V to 50 V. The I – V characteristics were measured in the dark and by irradiating the LT-GaAs working surface with a 150 W cold light source.

3. Results and Discussions

3.1. Experiment

Measurements of the dark current were carried out as a function of the bias voltage for a flag-type PCA with fired contacts at room temperature. The photocurrent was measured under irradiation of the LT-GaAs working surface with a 150 W cold light source.
The results of measuring the dark current and photocurrent PCA are summarized in the table for the purpose of comparative analysis. The photocurrent was determined without laser irradiation of the interelectrode gap. The contact firing procedure was not performed over sample No. 1.
The samples were annealed in a high-vacuum chamber at a temperature (Tann) of 580℃. The annealing time was 6 min in a vacuum of 10^{-7} Torr without stabilization by an arsenic flow, while the front side of the structure was tightly covered with a GaAs substrate. A sufficiently short annealing time was used to prevent a strong decrease in the density of arsenic clusters, as well as segregation of arsenic to the surface of the structure and its subsequent evaporation. Further, the contacts were fired in a high-vacuum chamber at a temperature of 410℃ for 2 minutes in a vacuum with a pressure of 10^{-7} Torr.

Table 1. Conditions for growth, annealing, and annealing of contacts for PCA, dark current, and photocurrent at room temperature.

Sample No.	Dark current, nA		Photocurrent, μA	
	U = - 50 V	U = + 50V	U = - 50 V	U = + 50V
1	- 30	16	-0.8	1.1
2	-50	38	-3.1	2.5
3	-40	25	-1	1 .3
4	-45	40	-1.7	2

As can be seen from Table 1, firing the contacts led to an increase in the photocurrent, on average, 2 times. The scatter in the values of the dark and photocurrent for samples (2 - 4) prepared under the same conditions is possibly due to defects in lithography, which manifests itself in a change in the distance between the electrodes. Embedding optimizes the distribution of the electric field within the structure of the photoconductive antenna. The main reason for the increase in the dark current and photocurrent in our structures after annealing is an increase in the electric field in the interelectrode gap due to the deepening of the contacts. This is explained as follows. In conventional photoconductive devices, the highest bias voltage that can be applied will be determined by the breakdown electric field of air, which is 2×10^6 V / m. In the proposed device, the greatest electric field is in the GaAs substrate, which has a higher electric breakdown field - 4×10^7 V / m. Therefore, with the same electrode gap width, the proposed device will withstand a higher bias voltage than conventional ones. This will lead to an even higher THz radiation power for the proposed device since it is obvious that the radiated THz power will increase with the applied bias.

3.2. Simulation

The PCA finite element simulation of an antenna with fired nanoscale contacts and conventional LT-GaAs protruding above the working surface has been carried out.

In the finite element method used, the Maxwell wave equation with the Poisson drift-diffusion equations are jointly solved, which makes it possible to use it to calculate the current-voltage characteristics in an LT-GaAs semiconductor with given values of the mobility of charge carriers, their lifetime, depending on the power laser radiation, parameters of a Gaussian beam. The electrophysical and optical parameters of modeling were used, which are shown in Table 2.

Table 2. Conditions for growth, annealing, and annealing of contacts for PCA, dark current, and photocurrent at room temperature.

Parameters	Values
LT-GaAs thickness	500 nm
LT-GaAs permittivity	12.9
LT-GaAs refractive index, real part	3.679
LT-GaAs refractive index, imaginary part	0.063

LT-GaAs band gap	1.42 eV
LT-GaAs affinity	4.07 eV
LT-GaAs carrier lifetime	Electron: 0.1, hole: 0.4 (ps)
LT-GaAs mobility	Electron: 400, hole: 100 ($cm^2 V^{-1} s^{-1}$)
N-type doping	10^{15} cm^3
LT-GaAs Auger recombination factor	7e-30[cm^6/s]
Temperature	300 K
DC bias voltage	0-50 V

In figure 4 shows the electric field through the gap of the photoconductive antenna at the electrode surface (Figure 2a) and at a depth of 80 nm (Figure 2b), where the electric field in the proposed model is the highest.

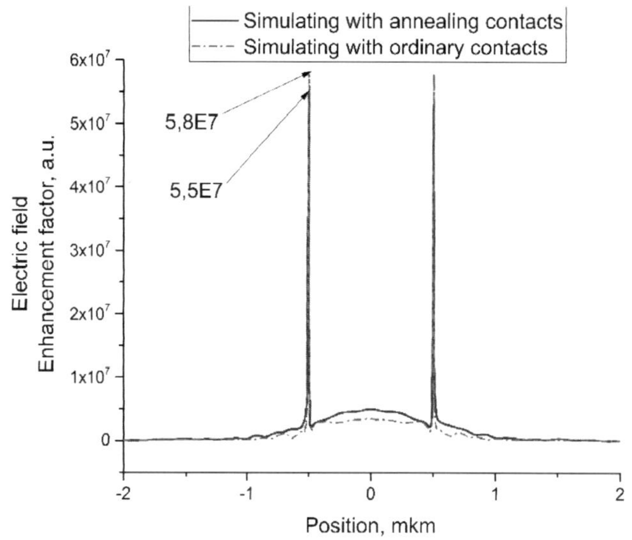

Figure 4. PCA electric field profile at a depth of 80 nm, with an electrode thickness of 80 nm.

It is assumed that the highest power of the THz response of the PCA of an antenna with nanosized electrodes can be expected in the case of a higher concentration of the electric field near the edge of the electrodes due to plasmon effects. This results in higher absorption of laser radiation when the antenna surface is irradiated.

In PCA with fusion contacts, the peak of the electric field strength near the electrodes is 5.5 * 10^7 a.u., which is slightly less than 5.8 * 10^7 a.u.. in comparison with conventional electrodes (Fig. 4), this is because there was a redistribution of the electric field in comparison with FPA without fired contacts - most of the laser radiation was absorbed in LT-GaAs. This can be seen in Fig. 5, which shows the distribution of the electric field in-depth, along the y-axis, along the direction to the antenna substrate.

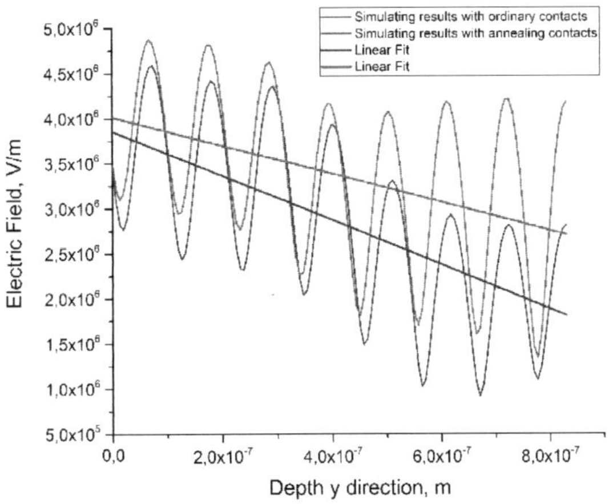

Figure 5. The electric field of the simulated PCA with embedded contacts (red line) and conventional PCA (blue line) at a bias voltage of 4 V. The obtained dependences were interpolated by a linear function.

The electric field calculated at the center of the antenna gap of both conventional and proposed photoconductive antennas is plotted against the depth of the substrate, starting from the surface of the substrate, as shown in Fig. 1. The electric field strength is 1.5 times greater for PCA with fired contacts, 80 nm thick than for conventional PCA.

Analysis of experimental data and simulation results shows that firing contacts increase the photocurrent of THz PCA, as seen from Fig. 6. In fig. 6 b) shows the simulation results, the dependence of the photocurrent on the bias voltage for PCA with fired contacts with a thickness of 80, 100, 200, 500 nm (IU80, IU100, IU200, IU500, respectively) and comparison with conventional PCA without fired contacts.

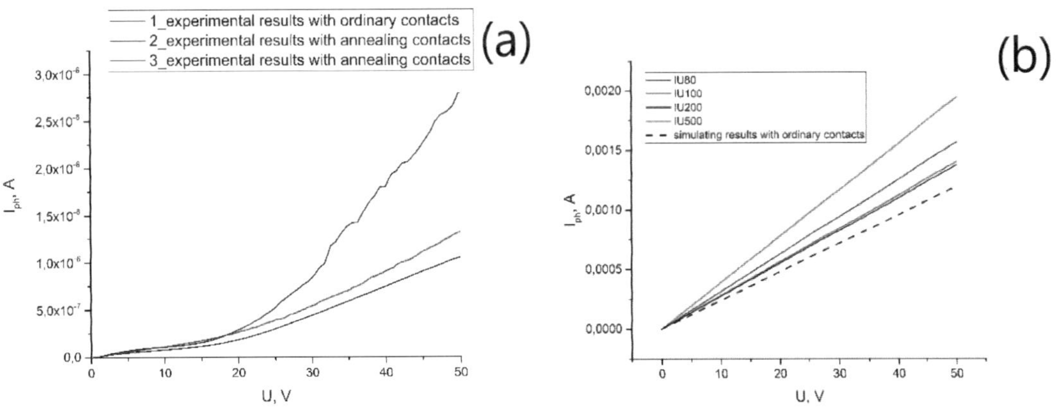

Figure 6. Dependence of the photocurrent on the bias voltage under optical laser irradiation with a wavelength of 800 nm, a) experimental data, the difference between conventional and embedded contacts; b) simulation results, a comparison was made of a conventional antenna with a contact thickness of 80 with a PCA with embedded contacts with different contact thicknesses: 80, 100, 200, 500 nm.

The difference between the experimental and theoretical values is because the photocurrent obtained experimentally was measured under irradiation with white light with an intensity much lower than that of a pump laser. The design of the electrodes with fired contacts makes it possible to assume that the expected THz power of the proposed one will also increase.

IOP Conf. Series: Materials Science and Engineering 1000 (2020) 012010 doi:10.1088/1757-899X/1000/1/012010

4. Conclusions

In this work, we simulate the PCA with a metal electrode height from 80 nm to 500 nm. The experiments carried out with the created PCA with embedded contacts demonstrated a twofold increase in the photocurrent as compared with conventional electrodes. Numerical simulation using the finite element method shows that the proposed structure of a THz antenna with fired electrodes leads to an increase in the photocurrent.

Simulation of the volt-ampere characteristics of the PCA under the influence of laser irradiation at a voltage bias of 50 V for an antenna with embedded contacts with a contact thickness of 80, 100, 200, 500 nm, in relation to a conventional PCA, showed an increase in the photocurrent generated by the antenna by 10, 12, 50 and 100%, respectively. Embedded contacts increase the electric field inside the low-temperature gallium arsenide.

5. Acknowledgments

This work was supported by the Competitiveness Program of National Research Nuclear University "MEPhI". Samples have been fabricated and partially investigated in Shared Research Center "Heterostructure microwave electronics and Physics of wideband semiconductors". This work was supported by Ministry of Science and Higher Education of the Russian Federation (project No. 0336-2019-0008).

References

[1] H. Roehle, R. J. B. Dietz, H. J. Hensel, J. Böttcher, H. Künzel, D. Stanze, M. Schell, and B. Sartorius, Opt. Express, **18,** 2296-2301 (2010).

[2] A.A.B. Sajak, Doctor of Philosophy thesis, University of Liverpool (2019). DOI: 10.17638/03031859.

[3] M. Bashirpour, S. Ghorbani, M. Kolahdouz, M. Neshat, M. Masnadi-Shirazi, H. Aghababa, RSC Adv., **7**, 53010–53017, (2017). DOI: https://doi.org/10.1039/C7RA11398F.

[4] D.V. Lavrukhin, R.R. Galiev, A.Y. Pavlov, *et al.,* Opt. Spectrosc. **126,** 580–586 (2019). DOI: https://doi.org/10.1134/S0030400X190501994.

[5] M. Tani, IEEE Photonics Technology Letters, **20**(12),1054-1056 (2008). DOI: 10.1109/LPT.2008.924184.

[6] J. Zhang, Design and performance of a terahertz photoconductive antenna with nano-crossfinger structure, Arxiv (2014) 1–8 http://www.opticsinfobase.org/ abstract.cfm?uri=col-12-8-083201%5Cnhttp://arxiv.org/abs/1408.0821.

Electrophysical properties of Janus-like silver nanoparticles

A Nomoev[1,2], **T Chimytov**[1,2]

[1]Institute of Material Physics of Siberian Branch of Russian Academy of Science, 6 Sakhyanovoi Street, Ulan-Ude, 670047, Russia
[2]Department of Physics and Engineering, Buryat State University, 24a Smolin Street, Ulan-Ude, 670000, Russia

E-mail: nomoevav@mail.ru

Abstract. The article presents the results of an experimental study of the electrophysical properties of Janus-like silver nanoparticles Ag/Si. The distribution of the electric potential of silver nanoparticles was obtained using atomic force microscopy using the Kelvin probe method. The electrical capacity of the particles is estimated.

Recently, in connection with the development of nanoelectronic technologies, there is a great interest in Janus-like nanoparticles. The presence of an intrinsic dipole moment and electrical capacity of such particles provides additional opportunities for increasing the efficiency of the interaction of a particle with external electric and alternating magnetic fields.
In the current paper, we study particles of Ag/Si nanopowder obtained using gas-phase synthesis [1]. The nanopowder comprises agglomerates of various sizes (Fig. 1), which are interconnected particles of silver and silicon. In fig. 1 on the left, it can be seen that the agglomerates under study comprise spherical particles (about 100 nm). Elemental analysis indicates that the largest proportion of silicon falls on the substrate (comprising silicon) and decreases in the agglomerate's bulk (Fig. 1 - in the middle). On the right in Fig. 1 - distribution density of silver, which is maximum in the agglomerates themselves.

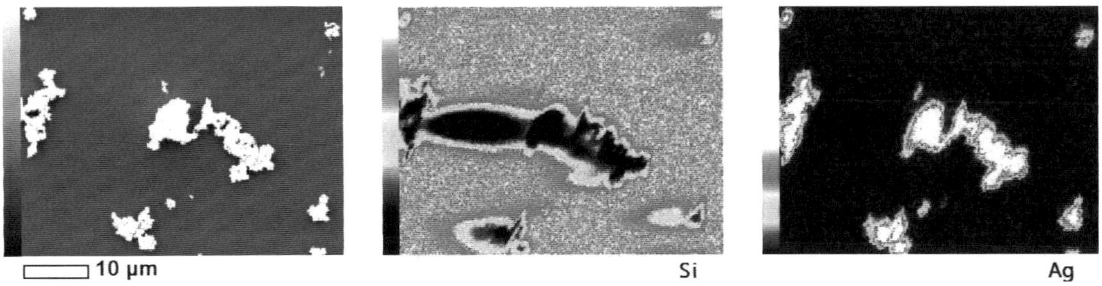

Figure 1. The chemical composition of agglomerates of Ag/Si nanoparticles, determined using a scanning electron microscope NEOSCOPE II JCM-6000

The different chemical nature of these elements leads to the formation of a Schottky barrier and the separation of electric charges within the particle itself, which leads to the appearance of a dipole moment. The work [2] provides an estimate of the dipole by using method based on measuring the dielectric constant. According to Langevin-Debye theory of molecular polarization the dielectric constant is related to the dipole moment p_0 as

IOP Conf. Series: Materials Science and Engineering 1000 (2020) 012011 doi:10.1088/1757-899X/1000/1/012011

$$\frac{(\varepsilon-1)M}{(\varepsilon+1)\rho}=\frac{1}{3\varepsilon_0}N_A(\alpha+\frac{p_0^2}{3kT}),$$

where α is the total polarization, which is measured experimentally. The dipole moment for Ag/Si nanoparticles is estimated at about $1{,}05\cdot10^{-29}$ C·m.

The distribution of surface potential on Ag/Si nanoparticles was measured using a SolverNext atomic force microscope (AFM) (NT-MDT, Russia). To study the contact potential difference of the particles, we used a HA_NC cantilever (TipsNano, Russia), with which samples were scanned in the Kelvin mode [3]. To exclude the influence of the surface relief, frequency and signal amplitude from the cantilever on the research results a two-pass technique is used. On the first pass of scanned line the topography of the sample is determined using semicontact mode with mechanical excitation of the cantilever. During the second pass, the distance between the scanned surface and the probe is kept constant. This distance should be large enough to exclude the influence of the topography. In this case, the probe is exposed only to long-range forces, the main contribution to which is made by the electrophysical properties of the nanoparticles. This dZ distance is set to provide sufficient signal level and lateral resolution. During this pass the cantilever is excited electrically by applying to the tip the voltage:

$$V=V_1+V_2\sin\omega t.$$

The frequency of the alternating electric field is set equal to the resonant frequency of the probe. The resulting capacitive force between the probe and surface which leads to the corresponding oscillations of the cantilever is

$$F_\omega=(V_1-\varphi)V_2\sin\omega t\frac{dC(z)}{dz},$$

where C is tip –surface capacitance. System feedback changes the variable component the probe potential V_{dc} while the ω component of the oscillations the cantilever do not disappear (when $V_{dc}=\varphi$). As a result the distribution V_{dc} reflects the distribution of surface potential over sample surface.

For AFM calibration, test samples were made in the form of polycore plates with partial platinum sputtering. A voltage of 3 V was applied to the platinum surface. Fig. 2 shows the potential distribution over the surface. It is clearly seen that the voltage varies in the range from 0 to 2.8 V.

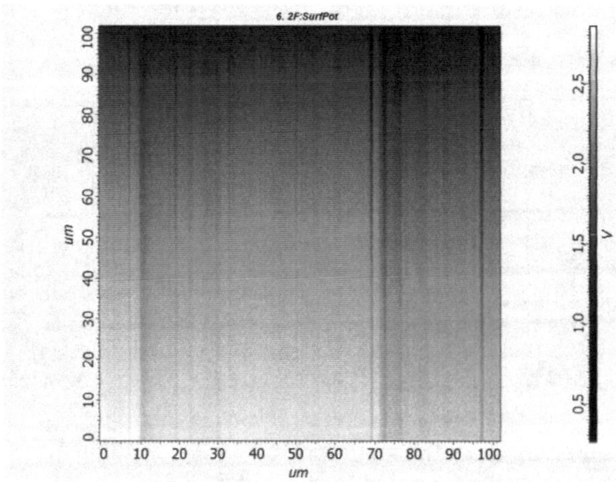

Figure 2. AFM image. The potential distribution of the test sample

The sample was a silicon substrate with particles of Ag/Si nanopowder deposited on it. Before preparing the samples, agglomerates of Ag/Si nanopowder particles were broken up by ultrasonic dispersion, by the method of acoustic cavitation in 99.7% isopropyl alcohol. For this, a cuvette containing a mixture of isopropyl alcohol and Ag/Si nanoparticles (mass fraction of powder - 0.5%)

was placed in a 50 W ultrasonic bath for 10 minutes. After that, the mixture was settled for 15–20 minutes, and the solution was applied to a silicon substrate using a pipette dispenser (DPOP-100-1000, Russia). After the solution has dried, the sample is suitable for AFM studies.
Fig. 3 shows the AFM scan results. The maximum potential difference for one particle is approximately 140 *mV*. The sign of the potential indicates that the particles are oriented by a metal contact to the surface of the silicon substrate. Knowing the value of the dipole moment we can estimate particle capacity of the order of 10^{-20} *F*.

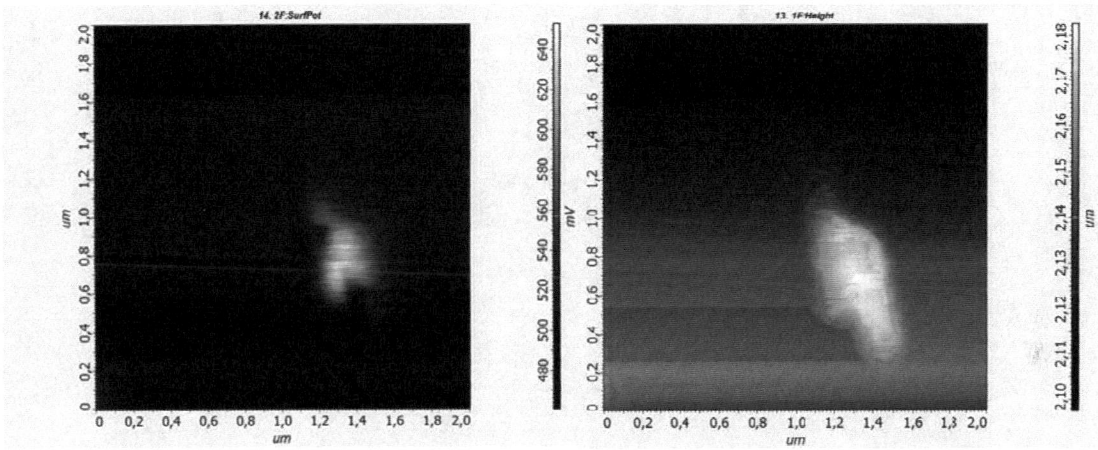

Figure 3. AFM image of an Ag/Si particle: potential distribution (left) and surface topography (right)

Conclusion

An experimental study of the electrophysical properties of silver nanoparticles has been carried out. The potential distribution over the surface deposited by nanoparticles was measured. The electrical capacity of the nanoparticle was estimated.

Acknowlegments

This work was financially supported by the project No. 0336-2019-0008 of the Russian Ministry of Education and Science and RFBR grant No.18-42-030004.

References

[1] Nomoev A, Bardahanov S 2012 *Technical Physics Letters* V 38 **8** 46–53
[2] Romanov N, Kalashnikov S, Tsyrenova M, Nomoev A 2015*Proc. Int. Conf. (Ulan-Ude)* **4** 132-136
[3] Nonnenmacher M, O’Boyle M P, and Wickramasinghe H K 1991 *Appl. Phys. Lett.* **58**, 2921

AUTHOR INDEX